GUIDEBOOK

Mineral Exploration, Mining and Aboriginal Community Engagement

Presented by:

In association with:

CBSR

See acknowledgements and the back cover for listing of our generous funding sponsors.

AUTHORS

Dan Jepsen, Executive Director & COO,
British Columbia & Yukon
Chamber of Mines

Bob Joseph, Principal,
Indigenous Corporate Training

Bill McIntosh, Principal,
Clean Communication

Bruce McKnight, Minerals Advisor

Mineral Exploration, Mining and Aboriginal Community Engagement: A Guidebook

First Edition, 2005
Mineral Exploration, Mining and Aboriginal Community Engagement: A Guidebook, 2005

Please forward inquiries to:
BC & Yukon Chamber of Mines
Suite 800 - 889 West Pender Street
Vancouver, BC V6C 3B2
Phone: 604-689-5271
Fax: 604-681-2363
Email: chamber@chamberofmines.bc.ca
www.chamberofmines.bc.ca
ISBN: 0-9738395-0-3

Printed on 100% Post Consumer Waste Recycled Paper

ACKNOWLEDGEMENTS

The founding partners of "*Mineral Exploration, Mining and Aboriginal Community Engagement (MEM-ACE): A Guidebook*" are pleased to present this work as a contribution to the development of mutually beneficial relationships between Aboriginal communities and the mineral exploration, mining, and resource sectors.

We are also pleased to acknowledge the financial support of the Guidebook's sponsors.

FOUNDING PARTNERS:

BC & Yukon Chamber of Mines (BCYCM)
Canadian Business for Social Responsibility (CBSR)

SPONSORS:

Alcan Inc.
Almaden Minerals Ltd.
Ashton Mining of Canada Inc.
Barrick Gold Corporation
BC Hydro
Blake, Cassels & Graydon LLP
Bull Housser & Tupper
Cameco Corporation
DeBeers Canada Inc.
Eagle Mapping Ltd.
Fraser Milner Casgrain LLP
Graymont Western Canada Inc.
Government of British Columbia
Indian and Northern Affairs Canada
Natural Resources Canada
Northern Trailer Ltd.
Pacifica Resources Ltd.
Placer Dome Inc.
Plutonic Power Corporation Inc.
Polaris Minerals Corporation
Prospectors and Developers Association of Canada
Redfern Resources Ltd.
Rimfire Minerals Corporation
Royal Bank of Canada
Syncrude Canada Ltd.
Talisman Energy Inc.
Teck Cominco Ltd.
Western Economic Diversification Canada
Western Keltic Mines Inc.
Yukon Zinc Corporation

Finally, we proudly recognize the many hours of effort, energy and commitment dedicated to this project by the following individuals through the BCYCM's First Nations and Community Relations Committee.

Without the support of these organizations and individuals, the production of the MEM-ACE Guidebook would not have been possible:

Brian Abraham,
Fraser Milner Casgrain LLP

Don Bragg, Prospector

Bob Carmichael, Redcorp Ventures Ltd.

Mike Casselman,
Candente Resource Corporation

Sue Craig, NovaGold Resources Inc.

Mike Farnsworth,
D.J.M. Farnsworth & Associates

Shari Gardiner, Hunter Dickinson Inc.

Michael Gray, Ultreia Capital Inc.

Dan Jepsen,
BC & Yukon Chamber of Mines

Dave Lefebure,
Ministry of Energy and Mines

Egil Livgard, Retired

Marlene Mathison,
BC & Yukon Chamber of Mines

Dean McDonald,
Committee Bay Resources Ltd.

Bruce McKnight, Minerals Advisor

Harlan Meade, Yukon Zinc Corporation

John Murray,
Murray Geological Services

Jamie Pardy,
Ministry of Energy and Mines

Robert Quartermain,
Silver Standard Resources Inc.

Leilah Tate, Independent Consultant

Randall Yip, Chartered Accountant

PREAMBLE

More than two years in the making, the Working Draft of the MEM-ACE Guidebook was first released in January, 2005 at "MINERAL EXPLORATION ROUNDUP 2005", the world's second largest mineral exploration conference. Hosted by BCYCM in Vancouver, B.C., Roundup 2005 was attended by over 5,000 delegates from 29 countries. In February, 2005, the Working Draft was presented at the Aboriginal Engagement and Sustainability Conference, hosted in Vancouver by CBSR, and drawing Presenters from as far away as New Zealand.

More than 700 copies of the Working Draft have been distributed, and input has been gathered from a broad range of Aboriginal, industry and government sources. That input has been carefully reviewed and represents a significant contribution to the current version of the MEM-ACE Guidebook. We are proud to present this work as the expression of considerable collaborative effort; we hope that it is received and used as a living document, to be enriched in future editions by broad discussion and feedback.

The MEM-ACE Guidebook is divided into two parts – "PART I: ABORIGINAL HISTORY AND BACKGROUND"; and "PART II: MINERAL EXPLORATION, MINING & ABORIGINAL COMMUNITY ENGAGEMENT – Case Studies, Pointers, Practical Advice and Best Practices." Designed and published as a companion piece to this Guidebook is "BUILDING SUSTAINABLE RELATIONSHIPS: 11 Case Studies and 4 Presentations Delivered at The Aboriginal Engagement & Sustainability Conference." (the AES Case Studies)

As the MEM-ACE Guidebook was born of British Columbia experience, it contains considerable material grounded in that context. We trust that readers in other jurisdictions will recognize that many of the principles developed from that experience are transferable, and we look forward to receiving feedback to help us to expand the perspective of future editions.

table of contents

Glossary of terms

Aboriginal Consultation The Crown has a legal duty to engage in meaningful consultation whenever it has reason to believe that its policies or actions, directly or indirectly, might infringe upon actual or claimed Aboriginal interests, rights or title. As the Supreme Court of Canada said recently, "The nature and scope of the duty of consultation will vary with the circumstances ... At all stages, good faith on both sides is required. The common thread on the Crown's part must be "the intention of substantially addressing [Aboriginal] concerns" as they are raised through a meaningful process of consultation ... Meaningful consultation may oblige the Crown to make changes to its proposed action based on information obtained through consultations ... The fact that third parties [industry] are under no duty to consult or accommodate Aboriginal concerns does not mean that they can never be liable to Aboriginal peoples. If they act negligently in circumstances where they owe Aboriginal peoples a duty of care, or if they breach contracts with Aboriginal peoples or deal with them dishonestly, they may be held legally liable. But they cannot be held liable for failing to discharge the Crown's duty to consult."[1]

Aboriginal Interest This is a broad term referring to the range of rights and entitlements that may arise from long use and occupation of traditional territories by Aboriginal people. Application of common law, statute law, treaty provisions, and the constitutional protection provided to "... the existing aboriginal and treaty rights of the Aboriginal people of Canada" by section 35 of *The Constitution Act, 1982,* to the facts of the particular case, determines the scope of "Aboriginal interest".

Aboriginal People These include Status Indians, Non-Status Indians, Métis and Inuit People.

Aboriginal Rights These are rights held by some Aboriginal peoples as a result of their ancestors' use and occupancy of Traditional Territories before contact with Europeans or before British sovereignty in Canada. Aboriginal Rights vary from group to group depending on what customs, practices, and traditions were integral to the distinctive culture of the groups.

Aboriginal Title In general, "Aboriginal title" refers to the rights of Aboriginal peoples to the occupation, use and enjoyment of their land and its resources. The classic legal definition was provided by the Supreme Court of Canada in Delgamuukw v. British Columbia[2]:

> "... aboriginal title encompasses the right to exclusive use and occupation of land; second, aboriginal title encompasses the right to choose to what uses land can be put, subject to the ultimate limit that those uses cannot destroy the ability of the land to sustain future generations of aboriginal peoples; and third, that lands held pursuant to aboriginal title have an inescapable economic component." (at para. 166, emphasis in original)

[1] *Haida Nation v. British Columbia (Minister of Forests)* 2004 SCC 73, Paragraphs 40-56

[2] [1997] 3 S.C.R. 1010

Band The *Indian Act* defines "band", in part, as a body of Indians for whose use and benefit in common, lands have been set apart. Each band has its own governing Band Council, usually consisting of a Chief and several councillors. The members of the band usually share common values, traditions and practices rooted in their language and ancestral heritage. Today, many bands prefer to be known as First Nations.

Band Chief Someone elected by band members to govern for a specified term.

Band Council or First Nation Council The band's governing body. Community members choose the Chief and councillors by election under section 74 of the *Indian Act,* or through traditional custom. The Band Council's powers vary with each band.

Elder A man or woman whose wisdom about spirituality, culture and life is recognized and affirmed by the community. Not all Elders are "old"; sometimes the spirit of the Creator chooses to imbue a young Aboriginal person. The Aboriginal community and individuals will normally seek the advice and assistance of Elders in a wide range of traditional and contemporary issues.

First Nation A term that came into common usage in the 1970s to replace the word "Indian", which many found offensive. The term "First Nation" has been adopted to replace the word "Band" in the name of many communities.

Hereditary Chief A Hereditary Chief is a leader who has power passed down from one generation to the next along blood lines or other cultural protocols, similar to European royalty.

Impacts and Benefits Agreement (IBA) A broad term used to describe various contractual commitments related to development of land or resources subject to Aboriginal rights. IBAs usually impose negotiated limits on a project's impacts on the environment, on fish and wildlife, on the land and First Nations' traditional use and enjoyment of same; and IBAs usually define a range of negotiated economic and preferential benefits to flow to the First Nation(s) whose lands are to be impacted by the development.

Indian The term "Indian" may have different meanings, depending on context. Under the *Indian Act*, Indian means "a person who pursuant to this Act is registered as an Indian or is entitled to be registered as an Indian". There are a number of terms employing the term "Indian" including Status Indian, Non-Status Indian and Treaty Indian. Status Indians are those who are registered as Indians under the *Indian Act*, although some would include those who, although not registered, are entitled to be registered. Non-Status Indians are those who lost their status or whose ancestors were never registered or lost their status under former or current provisions of the Indian Act. Treaty Indians are those members of a community whose ancestors signed a treaty with the Crown and as a result are entitled to treaty benefits. The term "Indian" was first used by Christopher Columbus in 1492 as he landed in the West Indies, believing he had reached India.

Indian Act The *Indian Act* is federal legislation that regulates Indians and reserves and sets out certain federal government powers and responsibilities toward First Nations and their reserved lands. The first *Indian Act* was passed in 1876, although there were a number of pre and post-Confederation enactments with respect to Indians and reserves prior to 1876. Since then, it has undergone numerous amendments, revisions and re-enactments. The Department of Indian Affairs and Northern Development administers the *Indian Act*.

Indigenous Peoples
a) "Peoples in independent countries whose social, cultural and economic conditions distinguish them from other sections of the national community, and whose status is regulated wholly or partially by their own customs or traditions or by special laws or regulations;"

b) "Peoples in independent countries who are regarded as indigenous on account of their descent from the populations which inhabited the country, or a geographical region to which the country belongs, at the time of conquest or colonization or the establishment of present state boundaries and who, irrespective of their legal status, retain some or all of their own social, economic, cultural and political institutions."[3]

Interim Measures Agreements (IMAs) The creation of a series of IMAs was one of the British Columbia Task Force recommendations that led to the current treaty process. IMAs are intended primarily to regulate the management and use of natural resources and lands – protecting and balancing the interests of the parties (including claimed Aboriginal title and rights) during the treaty negotiation process.

Inuit Aboriginal people in northern Canada, living mainly in Nunavut, Northwest Territories, northern Quebec and Labrador. Ontario has a very small Inuit population. The Inuit are not covered by the *Indian Act*. The federal government has entered into several major land claim settlements with the Inuit.

Métis People of mixed First Nation and European ancestry. The Métis history and culture draws on diverse ancestral origins such as Scottish, Irish, French, Ojibway and Cree.

Memorandum of Understanding (MOU) A formal agreement between governments or organizations. A relevant example is the 1993 MOU between the governments of Canada and British Columbia, allocating federal and provision responsibilities for funding various aspects of the current B.C. treaty-making process.

Reserves Lands set aside by the federal government for the use and benefit of a specific band or First Nation. The *Indian Act* provides that this land cannot be owned by individual band or First Nation members.

Socio-Economic Participation Agreement (SEPA) A synonym for Impacts and Benefits Agreement.

Traditional Ecological Knowledge (TEK) "TEK" broadly describes systems for understanding one's environment, based on detailed personal observation

[3] A number of definitions are in use; this widely recognized definition comes from ILO 169, a widely respected international convention.

and experience, and informed by generations of elders. TEK is recognized and used around the world as an important environmental assessment tool.

Traditional Territory Lands used and occupied by First Nations before European contact or the assertion of British sovereignty.

Treaty A formal agreement between the Crown and an Aboriginal people or peoples.

Treaty Rights Rights specified in a treaty. Rights to hunt and fish in traditional territory and to use and occupy reserves are typical treaty rights. This concept can have different meanings depending upon the context and perspective of the user.

Tribal Council A Tribal Council usually represents a group of bands to facilitate the administration and delivery of local services to their members.

PART I:

aboriginal
HISTORY AND BACKGROUND

Introduction

We hope to address two broad audiences and two broad purposes in PART I. Primarily, we are endeavouring to provide a foundation of Aboriginal history and perspective for the benefit of members of the mineral exploration and mining sectors. Historically, these sectors have not been well-prepared for engagement with Aboriginal communities and peoples. With the large majority of ownership, management and field staff coming from non-Aboriginal communities and cultures, the prevailing exploration and mining perspectives have been marked by assumptions and attitudes reflecting those of society-at-large. Until quite recently, non-Aboriginal society has not shown widespread understanding, respect, or tolerance for Aboriginal interests, rights, values, and culture.

We will not attempt to write a volume of history in these pages; our goal is to provide a meaningful outline of some important Aboriginal perspectives, revealing some of the historical roots for those perspectives, and calling for continued progress toward early, inclusive and open Aboriginal community engagement. While our primary audience will be the mineral exploration and mining sectors, we hope that this work will also reach a wider audience. Successful Aboriginal community engagement is an open process, requiring active commitment, mutual respect and informed communication on all sides.

Pre-Contact History

Until recently, archaeologists have generally agreed[4] that the ancestors of the Aboriginal peoples of North, Central and South America came from Asia some 12,000 years ago:

> "... via a land bridge known as Beringia, the land surrounding what is now the Bering Strait ... Archaeologists are certain of two things: that people have occupied this part of the Americas for at least 12,000 years, and perhaps for 20,000 years or more; and that migrations into these areas involved passing through portions of British Columbia."[5]

Generally, Aboriginal people are not impressed by archaeologists' land bridge theories regarding their origins and primal connection to their lands, as Nisga'a Chief David Mckay (Sim'oogit Axhlaawals) explained in 1888,

[4] Bob Joseph, President and Founder of INDIGENOUS CORPORATE TRAINING advises that recent research indicates that waves of migration have occurred over a longer period of time than indicated by the land bridge theory.

[5] Robert J. Muckle, *THE FIRST NATIONS OF BRITISH COLUMBIA*, (Vancouver: UBC Press, 1998) 15

> "I wish to say that every mountain and every stream has its name in our language, and every piece of country is known by the name our forefathers gave them ... God gave this land to our fathers a long time ago, and they made gardens and made homes and when they died they gave them to us."[6]

Aboriginal peoples' creation myths vary widely in their details, but a central theme links these origin stories. Generally, the people were placed in their territories by the Creator; then their enduring values, languages and customs were imparted – either by the Creator or other supernatural powers.[7] This sense of sacred trust and connection to the land underpins claims that Aboriginal title to ancestral lands was not extinguished by colonial occupation.

Nisga'a Chief James Gosnell (Sim'oogit Hleek) said in 1982,

> "What do we mean by aboriginal rights? In answering this question, the most important thing that must be borne in mind is that all of our rights flow from our relationship to the land. Our lives, our culture and our continued existence as a people are completely tied to the land in the area in which our ancestors have lived since time immemorial. That is why our people have indicated that we will never agree to an extinguishment of aboriginal title."[8]

By about 10,000 years ago, the glacial ice had retreated from most of British Columbia's land mass, allowing a wide variety of plant and animal life to establish, and permitting people to migrate from previously ice-free regions to the north, south, and southeast. Between 3,000 and 5,000 years ago saw the emergence of the varied and sophisticated "lifeways"[9] for which the Coastal Aboriginal peoples have become famous: i.e. primary reliance on salmon, complex social and political organization, large winter villages, advanced plant and stone technologies, and distinctive art. Similar (although less richly varied) development occurred in the southern interior during this period.[10]

6 "The Nisga'a have held the land of the Nass River in sacred trust from God since time immemorial. All Nisga'a culture is woven inextricably into this land. This is the story of their struggle to have this reality recognized by others and so have a meaningful and God-given place in Canada." From the Foreword, + Wii Ts 'imilx, Bishop of Caledonia. Editorial Committee of the Nisga'a Tribal Council, ***Nisga'a, People of the Mighty River***, (New Aiyansh: Nisga'a Tribal Council, 1992)

7 Muckle, ***THE FIRST NATIONS OF BRITISH COLUMBIA***, Ibid, 12

8 "***Nisga'a: People of the Mighty River***", Ibid.

9 "The description of traditional lifeways, often referred to as ethnography, is one of the basic goals of anthropology. In many places where First Nations did not have written forms of their languages at the time of European contact and where the cultures have undergone significant change since then, such as in British Columbia, anthropologists have relied primarily on oral traditions to reconstruct traditional lifeways. They recognize the potential problems in this, including deliberate attempts to deceive, but notwithstanding a few exceptions, most information gathered through oral tradition in British Columbia is considered reliable and is supported by archaeological and historical research." See Muckle, ***THE FIRST NATIONS OF BRITISH COLUMBIA***, Ibid, 26

10 "The unique, complex, and sophisticated cultures characterizing many First Nations of British Colombia for at least the last 3,000 years are often perceived as being directly related to the ability to harvest salmon in such abundance that it permitted the development of permanent or semi-permanent villages and sophisticated technology. With these changes came larger populations, social stratification, long-distance trade, warfare, heraldic art, and complex ceremonies such as the potlatch." Muckle, Supra, 17-18

Contact History

It is all but impossible for non-Aboriginal people to comprehend the enormity of the impact on the Indigenous peoples of the Americas created by the arrival of Christopher Columbus in 1492, and by the waves of Europeans who soon followed him.

Thomas R. Berger, former B.C. Supreme Court Justice and Commissioner of Mackenzie Valley Pipeline Inquiry of 1974-77[11], provides us with a powerful glimpse,

> "Since 1492, Native Institutions, their lifeways and their lands have been under attack. The history of the Americas has been the history of the encroachment of European societies on the Native peoples. The imposition of European values on Native communities has entailed an attempt to inculcate the European ideas of development, of economic progress, of the paramountcy of saving, accumulation and investment. Columbus, writing to the King and Queen of Spain, described the Indians of the West Indies as, 'So tractable, so peaceable ... that I swear to your Majesties that there is not in the world a better nation.' Nevertheless, he said, they should be 'made to work, sow and do all that is necessary to adopt our ways.'"[12]

Under the flag of Spain, the first Europeans visited the west coast of

what is now British Columbia in 1774. In eastern Canada, the English and French were extending their intermittent European conflict in a bitter struggle for control of New World peoples, territories, and trade. By contrast, the primary focus of the eighteenth century Spanish and Russian explorers who ranged up and down the west coast of North America was less invasive. Trading for furs and searching for an inland passage across the continent to the Atlantic mattered more than conquest. By the last decade of the 1700's, the Spanish had claimed the west coast from Mexico to Vancouver Island and the Russians had made an overlapping claim for control of the Pacific coast from Alaska to San Francisco. Neither nation had made much headway in establishing sovereignty.[13]

In 1778, Captain James Cook of Britain and his storm-battered crew became the first white people to land on the west coast of Vancouver Island, and to

[11] Berger's report of findings and recommendations, "Northern Homeland, Northern Frontier", is widely regarded as a seminal work in the struggle for jurisdictional justice among the world's northern peoples.

[12] Thomas R. Berger, *A LONG AND TERRIBLE SHADOW: White Values, Native Rights in the Americas, 1492-1992*, (Vancouver: Douglas & McIntyre, 1991), x-xi

[13] For an excellent discussion, see: Aniel Raunet, *WITHOUT SURRENDER, WITHOUT CONSENT: A History of the Nisga'a Land Claims*, (Vancouver/Toronto: Douglas & McIntyre, 1996, Second Edition) 17-25

chart the region.[14] The local Mohochat people called them "Mamahuit" – "the ones who lost their way in the fog."[15] By 1849 Britain had declared Vancouver Island as a Colony of the Empire; following in 1858 with the formation of the Colony of British Columbia. In general, we can say that European contact came much later and with far less initial impact in western Canada than in what became Canada's eastern provinces. Non-natives didn't begin to settle in British Columbia in significant numbers until 1858[16] - more than two hundred years after French settlement began along the banks of the St. Lawrence River.

Post-Contact History

On both the Atlantic and Pacific fronts, early developments suggested a more mutually beneficial approach to European-Aboriginal engagement than what ultimately transpired. In the initial years of contact, the balance of power was still in flux. That the European technological and military advantage was ultimately so overwhelmingly decisive makes it easy to forget how organized, sophisticated, and powerful many Aboriginal Nations were in those days. The Europeans desperately needed them as trading partners, wilderness-survival tutors, and military allies.

The Royal Proclamation of 1763

The pivotal British victory over France at the Battle of the Plains of Abraham in 1759 signalled the imminent collapse of New France. Montreal fell the following year. The famous *Royal Proclamation of 1763* was issued by Britain's King George III, mainly as an attempt to prevent conflict between the anticipated stampede of British settlers and the Aboriginal Nations. While it did not achieve its immediate purpose, the Proclamation had profound implications, both immediately and through the course of Canadian and American history.[17]

The *Royal Proclamation of 1763* declared that, it being

> "... essential to our Interest, and the Security of our Colonies, that the several Nations or Tribes of Indians with whom We are connected, and who live under our Protection, should not be molested or disturbed in the Possession of such parts of our Dominions and Territories as ... are preserved to them ... as their Hunting Grounds, therefore any lands that had not been ceded to or purchased by Us as aforesaid, are reserved to the said Indians. Furthermore, We do hereby strictly forbid, on Pain of our Displeasure, all our loving Subjects from making any

[14] This is the traditional view, but recently published evidence strongly suggests that in 1579, almost two hundred years earlier, Sir Francis Drake and his legendary Golden Hinde reached Vancouver Island (and perhaps as far north as the coast of Alaska) in the course of his three year, 65,000 km circumnavigation of the globe. In any event, there is no evidence of further European contact with British Columbia's west coast until the 1770's. See: Samuel Bawlf, ***The Secret Voyage of Sir Francis Drake***, (Toronto and Vancouver: Douglas & McIntyre, 2003)

[15] Aniel Raunet, ***WITHOUT SURRENDER, WITHOUT CONSENT***, Ibid, 18

[16] Muckle, ***THE FIRST NATIONS OF BRITISH COLUMBIA***, Ibid, 64

[17] The British policy of protecting Aboriginal lands from encroaching White settlers that underlay the Proclamation was one of the "Intolerable Acts" used to justify the American War of Independence. "The Declaration of Independence charged that the British King "... has excited domestic insurrection among us, and has endeavored to bring on the inhabitants of our frontiers, the merciless Indian Savages, whose known rule of warfare, is an undistinguished destruction of all ages, sexes, and conditions."

In the long term, the Proclamation would have other profound implications. It would, after the British had retired from North America, and, after two new nation-states, the United States and Canada, had been established, be instrumental in the recognition by the courts of both countries of the distinctive political status and aboriginal rights of the Indians." See Berger, ***A Long and Terrible Shadow***, Ibid, 62.

> Purchases or Settlements whatever, or taking Possession of any of the Lands above reserved, without our especial leave and Licence for that Purpose first obtained."[18]

The Douglas Purchase Treaties

James Douglas was the Governor of the new Colony of Vancouver Island from 1850 to 1864 (and of the mainland Colony of British Columbia from 1858 to 1864). Governor Douglas was a far-sighted and powerful administrator:

> "As the senior representative of the British Empire on the island, Douglas cultivated ties with the indigenous population. He believed that hostility toward natives would be a calamity for business, historian Robin Fisher writes in *Contact and Conflict.* Maintaining friendly relations with the natives, treating them equally under the law and limiting interference in their way of life, was in the interests of the Hudson's Bay Company.
>
> Douglas was prepared to assert the full force of his authority, as he demonstrated when he stood up to 200 armed natives at Cowichan in 1853. But as settlers replaced fur traders, igniting violent competition for land, he became a sympathetic advocate for the natives."[19]

Between 1850 and 1854, Douglas negotiated 14 separate purchase

treaties with Vancouver Island Aboriginal Nations.[20] The Douglas Purchase Treaties have legal and historical significance for two reasons. First, these 14 agreements were based on the principle of "cession" – a legal device approved by both British policy and international law at the time, for use by the Crown to acquire lands of

> "... a pre-existing society of indigenous people holding specific territories subject to cultivation. Acquisition would require the consent of the indigenous people to transfer their sovereignty and portions of all or a portion of their land to the acquiring state. This new relationship would then be set out in a formal treaty. Those acquiring the sovereignty and the territory were required to pay compensation to those who had ceded it."[21]

It would be argued later by other Aboriginal nations that the Douglas purchase treaties presumed and demonstrated British acknowledgement of pre-existing and unextinguished

18 Berger, ***A Long and Terrible Shadow***, Ibid, 61-62

19 Robert Matas, ***No 1: JAMES DOUGLAS - His mind was sharp, his hand was strong***, The Globe and Mail, April 16, 2005, S1. Matas noted: "James Douglas was a cold, crafty and selfish man with strong prejudices. He took ludicrous measures to appear grand, insisting on salutes being fired whenever he entered or left a fort. He was said to be a great hand at flogging. But without Old Square Toes, British Columbia would be part of the United States, the natives west of the Rockies would be nearing extinction and the development of the province would have been pushed back years, if not decades.
A Globe and Mail panel on the greatest British Columbian of all time had no doubts about who should be No. 1. Without hesitation, panel members Lily Chow, Daniel Francis, Jean Barman and Thomas Berger put Douglas at the top of the list."

20 Eleven of the treaties involved lands within today's Victoria, Sooke, and Saanich. Two involved Fort Rupert, and one was concluded at Nanaimo. These last three treaties were motivated by the Colony's need to secure coal deposits to power steam ships and to heat new homes and commercial buildings – representing the first non-agricultural economic activity embarked upon by non-Aboriginal people in British Columbia. The Douglas purchase treaties covered about 576 square kilometres, about three per cent of Vancouver Island's total land mass. See: Christopher McKee, ***TREATY TALKS IN BRITISH COLUMBIA: Negotiating a Mutually Beneficial Future***, (Vancouver/Toronto: UBC Press, 2000) 13

21 Supra, 14

Aboriginal title. Second, these 14 purchase treaties are important because, from the conclusion of the last of them in 1854, almost half a century would pass before British Columbia entered another treaty affecting Aboriginal land.[22]

Joseph Trutch and B.C.'s New Land Policy: Denial of Aboriginal Title

After James Douglas retired in 1864, British Columbia's policy toward Aboriginal rights and lands changed dramatically. Most of the reserves Douglas had established were reduced in size, and local authorities began to deny Aboriginal title in explicit terms.[23] Joseph Trutch, serving as the Commissioner of Land and Works, emerged as a major influence over the young colony's Aboriginal land policy.[24] His three major legacies were: to establish the British Columbia government's policy rejecting the existence of Aboriginal title, to reduce the size of the reserves established under Douglas' tenure as Governor, and to persuade the Colonial government that Aboriginal people should be prevented from using the legal device of "preemption" to obtain title to lands that were previously unoccupied.

On Aboriginal title, Trutch simply dismissed the Douglas purchase treaties as friendship agreements.[25] His policy position on Aboriginal title was made clear in an 1870 address to the Governor:

> "The title of the Indians in the fee of the public lands, or any portion thereof, is strictly denied. In no case has any special agreement been made with any of the tribes of the Mainland for the extinction of their claims of possession; but these claims have been held to have been fully satisfied by securing to each tribe, as the progress of settlement of the country seems to require, the use of sufficient tracts of land for their wants of agriculture and pastoral purposes."[26]

On the question of reserve size, Trutch's view was equally straightforward and convenient for the Colonial government and its non-Aboriginal citizens:

> "The Indians really have no right to the lands they claim, nor are they of any actual value or utility to them; I cannot see why they should either retain these lands to the prejudice of the general interests of the Colony, or be allowed to make a market of them either to Government or to individuals."[27]

22 Following an armed Aboriginal blockade that threatened the flood of Klondike-bound adventurers in 1898, the federal and provincial governments hastened to extend Alberta's Treaty 8 to cover Aboriginal lands in B.C.'s western Peace River Country.

23 The British Imperial Government's declining interest in the colony and in the welfare of its Aboriginal peoples was the main impetus for these policy changes. "Support among the English public for the general goals of the British Empire was on the decline, and both London and the Colonial Office saw little advantage in shouldering the costs of maintaining a significant interest in the colony. One result was that much of the control over Indian policy was left in the hands of local officials, who, while sharing interests in the land with other white settlers, were now dealing with issues that deeply affected their lives. It was not surprising, therefore, that Aboriginal peoples would receive less than fair treatment by the colonial government in the years to come." See: McKee, *TREATY TALKS IN BRITISH COLUMBIA*, Ibid, 17

24 Trutch's opportunity to influence government policy was enhanced by the merger of the Colonies of Vancouver Island and British Columbia in 1866.

25 Trutch argued that Governor Douglas had entered "... agreements with the various families of Indians ... for the relinquishment of their possessory claims in the district of the country around Fort Victoria, in consideration of certain blankets and other goods presented to them. But these presents were, as I understand, made for the purpose of securing friendly relations between those Indians and the settlement of Victoria, then in its infancy, and certainly not in acknowledgement of any general title of the Indians to the land they occupy." See: McKee, *TREATY TALKS IN BRITISH COLUMBIA*, Ibid, 18

26 Supra, 18

27 Supra, 19

Accordingly, Trutch ruled that reserve boundaries should be adjusted to restrict each Aboriginal family to a maximum of 10 acres each. A surveyor's report at the time noted that these "adjustments" released approximately 40,000 acres for white settlement.[28]

Finally, on the pre-emption question, Trutch persuaded the Colonial Legislature to pass the *Colonial Land Ordinance* on June 1, 1870. The Ordinance stated:

> "...any Male person being a British Subject, of the age of eighteen years or over, may acquire the right to pre-empt any tract of unoccupied, unsurveyed, and unreserved Crown Lands (not being an Indian Settlement) not exceeding Three Hundred and Twenty Acres in extent in that portion of the Colony situated to the Northward and Eastward of the Cascade or Coast Range of Mountains, and One Hundred and Sixty Acres in extent in the rest of the Colony.
>
> **Provided that such right of pre-emption shall not be held to extend to any of the Aborigines of this Continent,** except to such as shall have obtained the Governor's special permission in writing to that effect."

Essentially, this meant that a white settler could take title to an "unoccupied, unsurveyed, and unreserved" tract of land larger than a neighbouring reserve occupied by 30 Aboriginal families – who in turn were restricted to their reserve.

Viewed from a 21st century perspective, Trutch's views and measures may seem one-sided and draconian, but they resonated strongly in the expansionist political and economic climate of the day.[29] The 'Trutch View' endured as the core of the British Columbia government's Aboriginal policy for the next 125 years.

Disease, Death, and Depopulation

Throughout the Americas, Aboriginal contact with Europeans was soon followed with drastic declines in native population. With no natural immunity to communicable diseases introduced by the Europeans, and with little access to vaccines, Aboriginal people were decimated by waves of epidemics of smallpox, tuberculosis, scarlet fever, influenza, and measles.[30] British Columbia's mid-1700's Aboriginal

28 Supra, 19

29 "To many non-Aboriginal peoples, Trutch's views were persuasive because they were consistent with their views and with those of the local government. There was no better way to reinforce non-Aboriginal interests in the land than to demonstrate that Aboriginal peoples neither owned the land nor even conceived of owning it. Continuing Douglas's efforts at treaty-making was pointless: if Aboriginal title did not exist, there was no interest in the land that had to be purchased. Furthermore, the reserves set aside for Aboriginal peoples should not be seen as recognition of Aboriginal title or of the surrender of title to the land adjacent to reserves. Reserves were nothing more than gifts from the Crown to Aboriginal peoples." Supra., 19

30 "A group of ethnohistorians, led by Henry Dobyns ... asserts that epidemics were decisive demographic events. He estimates the population of North America before the arrival of Columbus to have been more than 18 million, and post-contact decline of 95 per cent over 130 years. He concludes that only the devastation of repeated epidemics could account for a loss of this magnitude.

In North America, Europeans looked upon the appalling losses among the Indian populations as providential. The depopulation of Native lands left large tracts available for colonization, to the satisfaction of many European immigrants. One Frenchman said that "it appears visibly that God wishes that they yield their place to new peoples."

Studies of the impact of the advent of Europeans in more recent times bear out the likelihood of a drastic and general decline in Indian numbers in North America owing to disease. In the 1890's, the American whaling fleet from San Francisco entered the Beaufort Sea and established whaling stations in the western Arctic. Eskimos were hired to gather driftwood to conserve the ships' stocks of coal, and to hunt caribou and muskoxen to supply the whalers with fresh meat. Whaling took a heavy tool of bowhead whales. But it was not just the animals that were affected by the coming of the whalers.

The whalers brought syphilis, measles and other diseases. When the whaling industry collapsed in 1908, of the original population of 2,500, there were only about 250 Mackenzie Eskimos left in the region between Barter Island and Bathurst Peninsula." See Berger, ***A LONG AND TERRIBLE SHADOW***, Ibid, Chapter 3, 33 population has been estimated at more than 250,000. By 1835, that number had fallen to about 100,000; by 1885, to 28,000; and by 1929, B.C.'s native population bottomed out at about 23,000.[31]

The Métis Nation

During the 1600's, France began establishing settlements along the St. Lawrence River to advance its fur trade interests. The prevailing "seigneurial" landholding system was a legacy of the feudal age: rigid and restrictive.[32] So, enterprising Frenchmen (soon to be known as "coureurs des bois" (runners in the woods) sought to improve their prospects by venturing into the wilderness, despite restrictions imposed on them by King Louis XIV. The coureurs des bois began establishing homes for themselves within or beside Indian communities, marrying Aboriginal women, and starting families. The birth of their children marked the birth of the Métis Nation.[33]

The Métis people played an important role in the history of Canada. They often acted as middlemen between Europeans and Aboriginals: encouraging trade and commerce, and mediating disputes. Also, and perhaps more significant, their legendary struggles for independence in the Red River Rebellion of 1869-70 and the North-West Rebellion of 1885 have inspired other Aboriginal peoples in their quests for greater autonomy, to the present day.

The Red River Settlement

In 1801, a group of Métis settled at the intersection of the Red and Assiniboine Rivers, where Winnipeg stands today. They were referred to as "Freemen", because they were bound by neither Indian nor Fur Trade company law. They set up narrow river lots similar to the seigneurial lots created earlier along the St. Lawrence.

In 1811, the Hudson's Bay Company granted 116,000 square miles of land in the fertile Red River Valley to Lord Selkirk. Efforts by the settlers to restrict Aboriginal hunting and trading practices led to conflict. In 1816, Cuthbert Grant Jr. led the Métis to victory at the battle of Seven Oaks, after which he unfurled the flag of the Métis Nation.

31 Muckle, *THE FIRST NATIONS OF BRITISH COLUMBIA*, Ibid, 60-61

32 Established in New France in 1627 and not officially abolished until 1854, the seigneurial system of land distribution and occupation relied on the personal dependency (i.e. indenture) of the "censitaires" (tenants) on the "seigneur" (landowner). In New France, the censitaires were called "habitants". See "Seigneurial System", HighBeam Research, www.highbeam.com/library/doc0/asp?

33 The word "Métis" comes from the Latin "Miscere" meaning "to mix" and is used generally to describe European men and Aboriginal women who had children together. The "coureurs des bois" and their Indian wives were the parents of the Métis Nation that began with the birth of their children. It is understood by Métis people today that the intermarriage of their ancestors involved more than just the blending of races and cultures; it was an evolution that culminated in the birth of a new Aboriginal Nation with its own language, called Michif. The Métis people had their own food, clothing, history, body politic, and flag. There is a national definition of Métis. It means a person who self-identifies as Métis, is of historic Métis Nation ancestry, is distinct from other Aboriginal peoples, and is accepted by the Métis Nation.

By the 1840's the Red River Settlement's population had swelled to 5,800 Métis and 1,600 non-Aboriginals. The Hudson's Bay Company was uneasy about the threat to its fur trading monopoly posed by the Settlement's burgeoning commercial activity. In 1849, the Company's Directors persuaded the North-West Police to charge Guillaume Sayer and three other Métis with "smuggling" furs. Louis Riel, Sr.[34] and 300 other Métis surrounded the courthouse and heard the traders pronounced "Guilty", but released without punishment. The Métis interpreted this ambiguous result as an acquittal and concluded that no one could be penalized for trading furs. The case is significant because it sparked the perception that the Hudson's Bay Company's fur trading monopoly had been broken, fanning the aspirations of the Métis and their allies.[35]

The new Dominion of Canada recognized the urgent need to consolidate its western and southern flanks against the threat of American expansion, and the Hudson's Bay Company saw that it was unable to control the increasingly powerful and independent Red River Settlers. So, in 1869, one of the key events in Canadian history quietly unfolded: The Dominion of Canada purchased the vast territory of Rupert's Land[36] from the Hudson's Bay Company for the sum of 300,000 English pounds. By that time more than 11,000 people lived in the thriving Red River Settlement; not one of them was consulted about the sale.

Led by Louis Riel, Jr., the Métis rebelled. After a period of brilliant military and political success, Riel blundered by allowing his new Provisional Government to execute Thomas Scott, an unrepentant Orangeman from Ontario who had been charged with bearing arms against the new state. Scott's instant martyrdom led to Riel's undoing. In 1871, the Parliament of Canada passed the Manitoba Act: providing for the acceptance of the Red River Settlement and surrounding territories into Canada as a full-fledged province, and calling for the dispatch of twelve hundred soldiers to 'protect' the settlements. As the troops from Ontario finally arrived, after more than three months of hard marching through the bush and muskeg, Riel belatedly recognized that they were coming to arrest him. He fled to the United States, where he remained in exile for more than a decade.[37]

[34] Louis Riel, Sr. was the father of Louis Riel, Jr., legendary leader of the Red River Rebellion of 1869-70 and the North-West Rebellion of 1885. The senior Riel also was a a man of uncommon resolve. He was known as the "Miller of the Seine. "To increase the flow of water over his mill wheel, the miller had dug a nine-mile channel from the Red River over to a creek that was named after the famous French river ... [Louis Riel, Sr.] had planned to build a woollen factory, but the mighty Hudson's Bay Company frowned on industry in fur country. See: Robert Hunter & Robert Calihoo, *OCCUPIED CANADA: A Young White Man Discovers His Unsuspected Past*, (Toronto: McLelland & Stewart, Inc., 1991) 84

[35] "For the next twenty years, the caravans of Red River carts continued to squeak and squeal back and forth unimpeded between the Red River and St. Paul [Minnesota]. Widespread trade in pots, kettles, stoves, farm implements and liquor grew year by year until, in 1867, it topped two thousand caravans a year." Ibid., 84

[36] The Company kept control of just 6 million of the 120 million acres stretching from Fort Garry to the Rocky Mountains.

[37] Riel remained a powerful political force in exile; he was elected three times to the House of Commons in absentia. Once he snuck into Ottawa to sign the Members' Register in the House, barely escaping with his life. Supra, 90

Louis Riel, Jr. and the Northwest Rebellion of 1885

The Northwest Rebellion of 1885 began with Louis Riel, Jr. returning with his wife, two small children and a few supporters to seize the church at Batoche, Saskatchewan, declaring "Rome has fallen." Riel formed the "Provisional Government of the Saskatchewan" and sent 400 men into action. Among the Métis' demands were local control of lands, responsible government, parliamentary representation, and confirmation of their land title in accordance with the river lot system survey. While the rebels won a few legendary guerrilla battles (i.e. at Duck Lake and Fish Creek, where the Métis and their predominantly Cree allies prevailed against odds of at least five to one)[38], in the end they were hopelessly out-manned and out-gunned.

In *OCCUPIED CANADA,* Robert Hunter[39] and Robert Calihoo provide a compelling account of the outcome (from a partisan perspective not often heard in non-Aboriginal Canada):

> "Carried from the Lakehead by train, the eight thousand-man North-West Field Force was deployed to staging points along the newly-laid tracks in time for a spring offensive ... When the inevitable siege of Batoche came, the Métis fought valiantly, with Riel crawling through the trenches mumbling prayers and Dumont [his trusted lieutenant] hissing at them to spare the ammunition. The advancing militia was equipped with 70,000 Gatling-gun rounds and 1.5 million bullets for their rifles. When, after four days, one of the [North-West] officers called for a bayonet charge, and led nine hundred screaming soldiers into Batoche to overcome two hundred out-of-work buffalo-hunters firing metal buttons, nails, and even pebbles, it was over swiftly. Riel's surrender three days later was almost as anticlimactic as the surrender of Poundmaker's Crees.
>
> In all about eighty whites and as many Métis and Indians had been killed. The Cree leaders, Poundmaker and Big Bear, were charged with treason-felony, even though Dewdney [Lieutenant Governor of the North-West Territories and Indian Commissioner] and his bosses in Ottawa knew that neither man had been guilty of insurrection. But these two were the remaining agitators for treaty revisions, and it was in Ottawa's interest to lock them away. The chiefs were each given three years in the newly built Stony Mountain Penitentiary, a gothic fortress in the middle of the plains north of Winnipeg. "I would rather prefer to be hung at once than to be in that place," Poundmaker cried. Prison life and prison guards broke the two proud chiefs: Poundmaker was so far gone physically after less than a year that he was released, only to die four months later.

[38] Supra, 117-122

[39] Robert Hunter was the first Chairman and President of the Greenpeace Foundation.

> Big Bear lasted longer; it took a little less than two years' imprisonment to destroy him. He was released in March 1887 and died ten months later."

As for Riel, he saw at least one of his visions fulfilled. Just before crossing into Canada on his way to Batoche, he had confided to a priest: "I see a gallows on top of that hill, and I am swinging from it." He was convicted of treason and hanged on November 16, 1885."[40]

CONSTITUTIONAL HISTORY OVERVIEW

The British North America Act

Canada officially became a country in 1867 with the passage of the *British North America Act.*[41] Pursuant to section 91(24) of the *Act*, the federal government was given authority to make laws about "Indians and lands reserved for the Indians".[42]

In 1869, Rupert's Land, including the Red River Settlement, was transferred by the Hudson's Bay Company to the Dominion of Canada. Louis Riel, Jr. blocked the Canadian delegation and Government surveyors from entering Rupert's Land. Riel's Métis provisional government issued the "Declaration of the People of Rupert's Land in the North-West."

The Indian Act

Many laws affecting Aboriginal peoples were combined in 1876 to become the *Indian Act*. The *Indian Act* gave Canada a coordinated approach to Indian policy rather than the pre-Confederation piece-meal approach.

There are three main areas of legislation contained under the *Indian Act*[43]: Land, Membership, and Local Government.

Assimilation Policy

By 1880, the Government of Canada had decided to apply its broad *Indian Act* powers to the strategic goal of assimilating the Aboriginal peoples into the mainstream of Canadian society. On May 5, 1880, Sir John A. Macdonald stood in the House of Commons to announce that his government's Indian policy was

> "... to wean them by slow degrees, from their nomadic habits, which have almost become an instinct, and by slow degrees absorb them or

40 Supra, 120-121

41 *Constitution Act, 1867* (U.K.), 30 & 31 Vict., c. 3 (formerly *British North America Act*, 1867)

42 It is important to recognize that while British Columbia's entry into Confederation in 1871 brought Section 91(24) into play, there were still many areas where the Province could (and did) use its jurisdiction to restrict Native ability to organize and act politically. For example, in 1872 British Columbia prohibited Aboriginal peoples from voting in provincial elections.

43 See: Hunter & Calihoo, *OCCUPIED CANADA*, Ibid, 143 for a strongly partisan Aboriginal perspective on the Indian Act: "In its initial efforts to deal with the problem of the surviving inhabitants of the land which Canada had just taken at gunpoint, the lawmakers in Ottawa cooked up a legislative witches' brew of regulations that covered every imaginable contingency in an Indian's life, leaving government agents hovering over his or her every activity from birth to death, with the power to snatch children from homes, monitor movements, prohibit "undesirable" activities, seize property, deny freedom of speech, religion, and self-expression, and throw "troublemakers" into jail promptly. Indians could not vote, drink, or own any land."
A less partisan but still very critical view of the *Indian Act* is provided by Stephen O'Neill: "The Indian Act is one of the most outmoded and archaic pieces of legislation in this country. This Act, first put on the statute books in 1876, clearly reflects a century-and-half-old view of the rights and responsibilities of Canada's Aboriginal peoples. The *Indian Act* is a sweeping code by which the federal government sets out rules and procedures affecting almost all aspects of life on an Indian reserve ... [T]he Minister of Indian Affairs is given the authority to make ultimate decisions on behalf of a band, either by approving decisions or overruling them." See: Stephen O'Neill, *DECISION MAKING ON RESERVES - THE CURRENT SITUATON*, in "Aboriginal Issues Today: A Legal and Business Guide", Ibid, 98

settle them on the land. Meantime they must be fairly protected."[44]

A number of specific measures were systematically deployed to give effect to this assimilation policy, as outlined below.

Women's Status

Since 1869, an Indian woman who married a non-Indian man lost her Indian status on marriage. The children of the marriage were also not entitled to Indian status. This provision was not changed until 1985, when the *Indian Act* was amended to remove discrimination against women, to be consistent with the *Canadian Charter of Rights and Freedoms*.[45]

Potlatch Law

From 1884 to 1951, the *Indian Act* prohibited Indians from participating in the Potlatch, the Sundance, and all other similar cultural ceremonies across Canada. To the west coast Aboriginal Nations, the Potlatch was an institution central to the culture, governance, and spiritual essence of the people, combining the processes of government, court, and church. Under the *Indian Act*, the election of Chiefs and Band. Council members replaced the Potlatch ceremonies for confirmation of Hereditary Chiefs.

Residential Schools

In 1844, the Bagot Commission of the United Province of Canada recommended training students in "as many manual labour or Industrial schools as possible … In such schools … isolated from the influence of their parents, pupils would imperceptibly acquire the manners, habits and customs of civilized life."[46]

In 1879, the Davin Report recommended residential schools based on the American model. Davin reported that the boarding school approach was the best answer because: it "… took [the Aboriginal child] from the reserve and kept him in the constant circle of civilization, assured attendance, removed him from the "retarding influence of his parents ...""[47]

Requiring Aboriginal children to participate in the public school system was seen as an important instrument of the assimilation policy, and compulsory attendance was incorporated into the *Indian Act* early in the twentieth

44 It should be noted that the seeds of this assimilation policy were sown decades earlier. In 1857, the legislatures of both Upper Canada and Lower Canada passed the *Gradual Civilization Act*, marking "… one of the most significant events in the evolution of Canadian Indian policy. Its premise was that by eventually removing all legal distinctions between Indians and non-Indians through the process of enfranchisement, it would be possible in time to absorb Indian people fully into colonial society.

… The new policy created an immediate political crisis in colonial/Indian relations in Canada. The formerly progressive and cooperative relationship between band councils and missionaries and humanitarian Indian agents broke down in acrimony and political action by Indians to see the act repealed. Indian peoples' refusal to comply and the government's refusal to rescind the policy showed that the nation-to-nation approach had been abandoned almost completely on the Crown side. Although it was reflected in subsequently negotiated treaties and land claims agreements, the Crown would not formally acknowledge the nation-to-nation relationship as an explicit policy goal again until the 1980's."
See: www.ainc-inac.gc.ca/ch/rcap/sg/sg23_e.html (The *GRADUAL CIVILIZATION ACT: Assimililating Civilized Indians*)

45 *Canadian Charter of Rights and Freedoms, Part I of the Constitution Act, 1982*, being Schedule B to the *Canada Act 1982* (U.K.), 1982, c. 11.

46 www.turtleisland.org/news/news-residential2.htm

47 www2.uoguelph.ca/dfischli/spotlights/s_p_davin.cfm "Report on Industrial Schools for Indians and Half-Breeds", Submitted in Ottawa on March 14, 1879.

century.[48] Hunter and Calihoo provide a bleak summary of the Aboriginal perspective on the result:

> "It began innocently enough. There was one thing the Indians all agreed they wanted: education. Promises were made about schools and teachers. The federal government could have funded day schools on the reserves or established boarding schools near the reserves, both of which might have stood a chance of working. Instead, it built three token "industrial schools," scattered at distant points from each other across the prairies, and otherwise tossed grant money into the laps of whichever churches wanted to take on responsibility for the education of Indian children. The fastest way to civilize the Indians, the priests and politicians agreed, was to remove the Indian kids from "the surroundings which tend to keep them in a state of degradation."
>
> Since it was widely assumed in the 1920's by the white man that the Indian didn't have the "physical, mental or moral get-up to enable him to compete," as one Indian Commissioner phrased it in a report to Ottawa, the only thing to do was to teach him a trade. Taking their grant money, the priests quickly set up residential schools – as far as possible from the bad influence of home, family, and friends. Indian children shared the common trauma of being dragged away from their homes and cast into strange places, where they were beaten if they got caught speaking their own languages ... Since the residential schools were all denominational, Christian indoctrination was the main focus: learning a trade came a poor second.
>
> Not surprisingly, children ran away from the residential schools in droves. So many escaped that the RCMP were used to chase them down and haul them back into classes. The "problem" of escaping children got to the point where the government passed a law stating that Indian parents had no authority over their children while the kids were in residential school.
>
> The situation would have been enough of a nightmare for parents, grandparents, aunts, uncles, and children alike, even if it hadn't turned out that the white men had built such sub-standard buildings into which to herd a captive generation of young Indians that many of the children sickened and died. In 1914, Duncan Campbell Scott, former Deputy Superintendent General of Indian Affairs confessed:
>
> "...the system was open to criticism. Insufficient care was exercised in the admission of children to the schools. The well-known predisposition of Indians to tuberculosis resulted in a very large percentage of deaths among the pupils. They were housed in buildings not carefully designed for school purposes, and these

[48] Nathalie des Rosiers, President, Law Commission of Canada, Remarks for a speech delivered at the "Moving Forward Conference", Sydney, Australia, August 15, 2001: "What distinguishes residential schools for Aboriginal children is that they were part of a policy of assimilation that was sustained for many decades: the residential school experience influenced the lives of several generations of people ... The Commission's review of the increasing amount of information on residential schools for Aboriginal children has led it to three conclusions. First, racial attitudes about the backwardness and inferiority of Aboriginal peoples fuelled the maltreatment and abuse experienced by children at residential schools. Second, the affronts to the collective dignity, self-respect and identity of Aboriginal peoples that occurred in residential schools are closely linked to the nature and scope of the redress individuals and communities now seek. Third, there remains today a significant need for public education. All Canadians must be offered the opportunity to understand the destructive influence of the residential school system and to appreciate why the federal government is morally obliged to take significant steps to help survivors and their communities."

> buildings became infected and dangerous to the inmates. **It is quite within the mark to say that fifty per cent of the children who passed through these schools did not live to benefit from the education which they had received therein.**""[49]

Many of today's Aboriginal leaders are survivors of the residential school program.[50] By 1931, 80 residential schools existed across Canada. In 1951, the federal government began a four-decade long process of shutting down the schools; the last residential school for Aboriginal children closed in 1986.

Land Claims and Aboriginal Protests

From an early stage of non-Aboriginal settlement, Aboriginal peoples understood the importance of political action and protest. In 1872, a group of Coast Salish chiefs rallied in front of B.C.'s provincial land registry office to demand the enlargement of reserves. In 1887, the chiefs of the Nisga'a and Tsimshian met with B.C. Premier William Smithe, demanding more reserve lands, treaties, and self-government. The Premier's views on land ownership, Aboriginal title, and assimilation made it clear that negotiation wasn't likely to be fruitful any time soon:

> "The land belongs to the Queen... A reserve is given to each tribe, and they are not required to pay for it. It is the Queen's land just the same, but the Queen gives it to her Indian children because they do not know so well to make their own living the same as the white man and special indulgence is extended to them, and special care shown. Thus, instead of being treated as a white man, the Indian is treated better. But it is the hope of everybody that in a little while the Indians will be so far advanced as to be the same as a white man in every respect."[51]

In 1913, the Nisga'a travelled to London to petition the Privy Council to settle their Land Question. In 1916, various interior and coastal Aboriginal communities formed B.C.'s first province-wide Aboriginal political organization – the Allied Tribes of British Columbia (ATBC). It seemed evident to ATBC that their chances of progress were much better through litigation than through more fruitless dealings with politicians and officials.[52]

By 1927, the Government of Canada had had enough, and it amended the *Indian Act* to make it illegal for Indians to retain a lawyer or raise money to advance their claims to land. This provision remained in effect from 1927 to 1951: denying Canada's Aboriginal peoples fundamental rights enjoyed by every other citizen.

49 Hunter & Calihoo, *OCCUPIED CANADA*, Ibid, 230-231

50 Phil Fontaine, National Chief of the Assembly of First Nations, was abused in two residential schools. See: Roy MacGregor, "This Country", The Globe and Mail, April 4, 2005, A2

51 Christopher McKee, *TREATY TALKS IN BRITISH COLUMBIA*, Ibid, 24

52 That view became even stronger in 1921, when the Judicial Committee of the Privy Council (then the highest Court in the British Commonwealth) ruled in a Nigerian case that "Aboriginal title was a pre-existing right that must be presumed to survive, unless established otherwise by the context or the circumstances in which the right operated." Ibid, 25

Other Legislation

Other federal and provincial legislation has played a role in creating the complex Aboriginal issues that challenge the current generation of Canadians.

For example, did you know?

- Aboriginal Peoples were prohibited from voting in provincial elections until 1948.
- Aboriginal Peoples were prohibited from voting in municipal elections until 1949.
- Aboriginal Peoples were prohibited from voting in federal elections until 1960.
- Aboriginal veterans of war lost their Indian status. Many Aboriginal World War II veterans found that when they returned home after fighting overseas for Canada, they were no longer considered Indians. The *Indian Act* specified that Indians absent from reserve for four years were no longer Indians.

Push for *Indian Act* Reform and Greater Aboriginal Control

In recent decades there has been significant pressure to address historical *Indian Act* issues. That pressure has been imposed by the combination of increasingly organized and effective political actions by various Aboriginal groups, and a series of Supreme Court of Canada decisions that established the validity of the concepts of unextinguished Aboriginal title and Aboriginal rights of self-determination.

Band Chiefs and Councils, other Aboriginal leaders across the country, and human rights leaders have called for increased Aboriginal autonomy from the federal Department of Indian Affairs and the *Indian Act*.

The White Paper of 1969

In 1969 the government of Canada introduced a White Paper on Indian Policy[53], calling for "the eventual elimination of the various "privileges" of Aboriginal peoples, with the ultimate goal of "normalizing" their integration into Canadian society. Among other things, the White Paper would have done away with the *Indian Act*, phased out federal obligations to Aboriginal peoples, and redistributed reserve lands based on individual ownership."[54] Confronted by Aboriginal opposition to the White Paper from coast to coast, the federal

53 See: www.ainc-inac.gc.ca/ch/rcap/sg/sg18_e.html

54 Christopher McKee, *TREATY TALKS IN BRITISH COLUMBIA*, Ibid, 27

government withdrew the initiative in 1971, replacing it with the "Core Funding Program" – supplying Aboriginal groups with resources to promote their causes through research, publication and legal action.[55]

Two years later came a commitment by the Minister of Indian Affairs, Jean Chrétien, to begin negotiations with the Nisga'a and to begin the process of northern treaty-making.

Canada's Constitution Act

In 1982 the Government of Canada patriated the Canadian constitution, and in so doing, formally entrenched Aboriginal and treaty rights in the supreme law of Canada.

Section 35 of the *Constitution Act, 1982* provides:

> "35(1) The existing aboriginal and treaty rights of the aboriginal people in Canada are hereby recognized and affirmed.
> (2) In this Act, "Aboriginal Peoples of Canada "includes the Indian, Inuit, and Métis Peoples of Canada.
> (3) For greater certainty, in subsection (1), "treaty rights" includes rights that now exist by way of land claims agreements or may be so acquired.
> (4) Notwithstanding any other provision of this act, the aboriginal and treaty rights referred to in subsection (1) are guaranteed equally to male and female persons."[56]

Section 35 neither confirms nor creates absolute Aboriginal rights. Before 1982, the Supreme Court had confirmed that Aboriginal rights could be extinguished in three ways: by sale or surrender of the right, by otherwise valid legislation that clearly and deliberately extinguishes an Aboriginal right, or by legislation that can only be interpreted as extinguishing an Aboriginal right. So, section 35(1)'s "existing Aboriginal or treaty rights" are rights that were not extinguished by surrender or legislation before 1982.

On the other hand, Aboriginal and treaty rights existing after proclamation of the *Constitution Act, 1982* now receive significant legal protection under section 35. Existing Aboriginal land rights can no longer be extinguished without the consent of those Aboriginal peoples holding interests in those lands. Aboriginal consent may be required to give effect to legislation purporting to extinguish Aboriginal land rights, even if compensation is paid. Finally, government regulation of Aboriginal land rights may still be possible, if appropriate and meaningful consultation is undertaken with the affected Aboriginal communities.[57]

55 Supra, 27

56 *Constitution Act, 1982*, being Schedule B to the *Canada Act 1982* (U.K.), 1982, c. 11.

57 *ABORIGINAL ISSUES TODAY: A Legal and Business Guide*, Edited by Stephen B. Smart and Michael Coyle, (North Vancouver, Self-Counsel Press, 1997) 46-47

Major Court Cases: The Supreme Court Leads the Way

Introduction

Beginning with the ground-breaking *Calder* case in 1973, the Supreme Court of Canada has used a series of important decisions to define and explain Aboriginal rights and title in Canada. Even more important than the technical rulings themselves, the Supreme Court's reasoning in this line of cases has established the validity of the concept of unextinguished Aboriginal title, has induced both the federal and provincial governments to undertake meaningful treaty-making, and has given Aboriginal peoples a much strengthened opening position for negotiations.

The decisions reviewed below underscore the need for all levels of government to avoid prolonged, expensive and divisive legal battles: i.e. to resolve claims of Aboriginal title and rights through negotiation rather than litigation.

Calder[58], Supreme Court of Canada, 1973

In the *Calder* case, the Nisga'a Tribal Council asked the courts to support their claim that Aboriginal title had never been extinguished in the Nass Valley, near Prince Rupert. Although the Supreme Court of Canada ultimately ruled against the Nisga'a on a technicality, the case is historic because for the first time, Canada's highest court ruled that Aboriginal title was rooted in the "long-time occupation, possession and use" of traditional territories. As such, Aboriginal title existed at the time of original contact with Europeans, and at the time of formal assertion of British sovereignty in 1846.

Shortly after the *Calder* decision[59], the Canadian government agreed to begin negotiating with the Nisga'a on their "Land Question"[60], and with northern Aboriginal peoples on treaties to define their rights to land and resources.

Guerin[61], Supreme Court of Canada, 1984

In *Guerin*, the judgment of Chief Justice Brian Dickson extended *Calder* to describe Aboriginal interest in land as a "pre-existing legal right not created by the *Royal Proclamation* ... the *Indian Act*...or any other executive order or legislative provision."

In 1955, the Musqueam First Nation approved a surrender "in trust" of some of its reserve land in the city of Vancouver, for the purpose of a lease to the Shaughnessy Golf and Country Club. The lease transaction had been discussed in detail with the Band and

58 *Calder v. Attorney-General of British Columbia*, [1973] S.C.R. 313

59 Prime Minister Pierre Trudeau met with Frank Calder in the months following the release of the historic 1973 decision of the Supreme Court in the *Calder* case, and remarked "that the case [had] pushed him to reconsider the colonialist assumptions underlying his administration's policy on Aboriginal peoples and to acknowledge the possibility of Aboriginal self-determination, treaty rights, and self-government as key organizing principles." McKee, *TREATY TALKS IN BRITISH COLUMBIA*, Ibid, 27-28

60 See: *Nisga'a: People of the Mighty River*, Ibid, 1

61 *Guerin v. The Queen*, [1984] 2 S.C.R. 335.

Band consent had been given based on those discussions. The Crown subsequently concluded a lease on terms substantially different and less advantageous to the Musqueam. The true terms of the lease were not disclosed to the Band until 1970.

The Supreme Court ruled that the federal government had a "fiduciary responsibility" for Aboriginal people – that is, a responsibility to safeguard Aboriginal interests. This duty placed the government under a legal obligation to manage Aboriginal lands as a prudent person would when dealing with his/her own property. The Court held the government had breached this fiduciary duty and awarded damages of $10 million to the Musqueam.[62]

The *Guerin* ruling is significant because it recognized pre-existing Aboriginal rights both on and off-reserve.

Martin[63], Supreme Court & Court of Appeal of British Columbia, 1985 - ongoing (the Meares Island Case)

In 1984, the Nuu-chah-nulth people and other protesters blocked MacMillan Bloedel's access to its timber berth on Meares Island. The Province of British Columbia regarded the vast majority of the island as Crown land, but the

protesters claimed that allowing logging on Meares Island interfered with Aboriginal title. A court injunction was sought to halt MacMillan Bloedel's operations until the claim was resolved.

The B.C. Supreme Court denied the request, but the B.C. Court of Appeal (which does not usually grant leave to hear appeals in such injunction cases)[64] agreed to hear the application. In granting the injunction in a three to two decision, two judgements for the majority gave some pointed feedback to the provincial Crown. Justice Seaton said:

> "It has ... been suggested that a decision favourable to the Indians will cast a huge doubt on the tenure that is the basis for the huge investment that has been made and is being made... **There is a problem about tenure that has not been attended to in the past. We are being asked to ignore the problem as [the province of British Columbia**

[62] "Many First Nations that have received less than fair market value have since filed claims for compensation from the Crown. Other First Nations have surrendered lands to the Crown for sale; the lands have not been sold by the Crown, leaving these First Nations without compensation. These lands are treated as provincial Crown lands, because under Canadian law, the surrender gave full ownership to the province. The principles set out in the *Guerin* decision are frequently relied on to support claims for the return of these lands." Smart and Coyle, *ABORIGINAL ISSUES TODAY*, Ibid, 45

[63] *MacMillan Bloedel Ltd. v. Mullin; Martin v. R. in Right of B.C.* [1985], 61 B.C.L.R. 145, (B.C.C.A.).

[64] The key issue for the Nuu-chah-nulth was their desire to preserve evidence of their historic use of the natural resources of the area - they argued that clear-cut logging would erase that evidence.

has] ignored it. I am not willing to do that."[65]

Justice MacFarlane was even blunter in calling for meaningful treaty-making negotiations:

> "The fact that there is an issue between the Indians and the province based on Aboriginal claims should not come as a surprise to anyone. Those claims have been advanced by the Indians for many years. They were advanced in [the *Calder* case] and half the court thought they had some substance ... **I think it is fair to say that, in the end, the public anticipates that the claims will be resolved by negotiations and by settlement. This judicial proceeding is but a small part of the whole process which will ultimately find its solutions in a reasonable exchange between governments and the Indian nations.**"[66]

The *Martin (Meares Island)* case was adjourned by agreement of the Nuu-chah-nulth First Nation, MacMillan Bloedel, and the governments of British Columbia and Canada. The injunction on logging is still in effect and none of the parties has requested resumption of the trial. As a result of *Martin*, B.C.'s provincial Ministry of Native Affairs was created in 1988.[67] By 1989, public support for the government's entry into treaty negotiations had reached 80 per cent, and Social Credit Premier Vander Zalm appointed a Native Affairs Advisory Committee to consider the government's options. By the fall of 1990, following the Mohawk blockades at Oka and Kahnawake, and further Aboriginal blockades across the country (particularly in Alberta and B.C.), Premier Vander Zalm announced that his government would commence negotiations with B.C.'s Aboriginal and First Nations (still without acknowledging Aboriginal title).[68]

Sparrow[69], Supreme Court of Canada, 1990

In the *Sparrow* case, a member of the Musqueam First Nation appealed his conviction on a charge of fishing with a longer drift net than permitted by the terms of the Band's fishing license under the *Fisheries Act*. He based his appeal on the argument that the restriction on net length was invalid because it was inconsistent with Section 35 of the *Constitution Act, 1982* – the section of the *Act* that recognizes and affirms existing Aboriginal and treaty rights.

The *Sparrow* case was the first opportunity for the Supreme Court of Canada to interpret what Section 35 actually meant. In overturning Sparrow's conviction, the Court ruled that the *Constitution Act* provides "a strong measure of protection" for Aboriginal rights, and that any

65 *Martin*, Ibid, 160 (BCCA)

66 Supra, 172-173

67 Frank Calder was named B.C.'s Minister of Native Affairs, becoming Canada's first Aboriginal cabinet minister. See: www.nisgaalisims.ca/pdf/June04ExecSummary.pdf

68 Christopher McKee, *TREATY TALKS IN BRITISH COLUMBIA*, Ibid, 30

69 *R. v. Sparrow* [1990] S.C.R. 1075.

proposed government regulations that infringe on the exercise of those rights must be constitutionally justified. The two-part Sparrow test for determining whether an infringement can be justified is:

i) the government must be acting pursuant to a valid legislative object;
ii) and the government's actions must be consistent with its fiduciary duty toward Aboriginal peoples.

If a valid legislative object is established, assessment of whether the government's actions are consistent with that fiduciary duty between the Crown and Aboriginal peoples requires that three questions be addressed:

i) Has there been as "little infringement as possible" in order to achieve the intended result?
ii) In a case of expropriation, has fair compensation been paid?
iii) Has the particular Aboriginal People been consulted?[70]

The Sparrow justification test applies beyond Aboriginal rights, to include treaty rights and Aboriginal title as well.[71]

The Sparrow Court further ruled that:

- Aboriginal and treaty rights are capable of evolving over time and must be interpreted in a generous and liberal manner
- Governments may regulate existing Aboriginal rights only for a compelling and substantial objective, such as the conservation and management of resources
- After conservation goals are met, Aboriginal people must be given priority to fish for food over other user groups.

Delgamuukw[72], B.C. Supreme Court & Court of Appeal, Supreme Court of Canada, 1984-1997

The *Delgamuukw* cases are critical pieces of the constitutional puzzle of Aboriginal rights and title for British Columbia and all of Canada. Together, the three *Delgamuukw* decisions summarized below raise a number of important issues addressed in the recently released Supreme Court of Canada decision in *Haida Nation*[73], and provide a good foundation for our review of that case.

In 1984, 35 Gitxsan and 13 Wet'suwet'en Hereditary Chiefs asked the Supreme Court of British Columbia to recognize their ownership of 57,000 square kilometres of land in north-western B.C., to confirm their right to govern their traditional territories, and to award compensation for loss of their lands and resources. The Gitxsan and Wet'suwet'en elected to proceed with trial by judge alone (rather than by judge and jury) and submitted an enormous body of oral and written evidence (the Court transcript covered 369 days of

70 Sparrow, Ibid, 1119

71 For a good summary and analysis of Sparrow and related cases, see: Charles F. Willms and Alison Kearns (articled student), *The Haida Nation v. Weyerhaeuser: Upcoming Supreme Court Decision and Its Implications for Aboriginal-Energy Partnerships and Joint Ventures*, (FASKEN MARTINEAU) www.fasken.com

72 Delgamuukw v. British Columbia [1997] 3 S.C.R. 1010

73 Haida Nation v. British Columbia (Minister of Forests, Neutral Citation: 2004 SCC 73

proceedings) regarding the nature and duration of their use and occupation of their traditional lands. Counsel for the Gitxsan and Wet'suwet'en presented a legal argument that has been described as "the most extensive ever mounted in an Indigenous rights case."[74]

In his Reasons for Judgment released in 1991, Chief Justice McEachern left open the possibility that Aboriginal rights may arise through the use and occupation of specific lands for Aboriginal purposes for an indefinite (and lengthy) period prior to British sovereignty. However, he ruled that in any event the Crown had extinguished any such Aboriginal rights by its imposition of complete dominion over the Colonial territory prior to joining Confederation in 1871.[75] The Gitxsan and Wet'suwet'en appealed.

In 1993, the B.C. Court of Appeal reversed much of the lower court's decision and ruled instead that the Gitxsan and Wet'suwet'en peoples do have "unextinguished non-exclusive Aboriginal rights, other than a right of ownership," to much of their traditional territory. In addition, the appeal court Justices strongly recommended that the scope and content of those rights would best be defined through negotiation rather than litigation. British Columbia appealed to the Supreme Court of Canada.
On December 11, 1997, a unanimous Supreme Court handed down its landmark *Delgamuukw* ruling, providing some important definition and description of Aboriginal title, confirming the legal validity of Aboriginal oral history, and clarifying the nature of the Crown's duties of consultation and accommodation in the context of infringement of Aboriginal rights.

Aboriginal Title

Chief Justice Lamer's judgment for the Court provided the most comprehensive Supreme Court description of Aboriginal title seen to date:

> "[A]boriginal title encompasses the right to exclusive use and occupation of the land held pursuant to that title. (at para. 117) What aboriginal title confers in the right to the land itself. (at para. 138) [A]boriginal title encompasses the right to exclusive use and occupation of land; second, aboriginal title encompasses the right to choose to what uses land can be put, subject to the ultimate limit that those uses cannot destroy the ability of the land to sustain future generations of aboriginal people; and third, that lands held pursuant to aboriginal title have an inescapable economic component." (at para. 166, emphasis in original)

Justice Lamer pointedly outlined the double-edged implication of that "inescapable economic component": on the one hand, it means that the

74 Michael Jackson, *A Legal Overview - The Case in Context*, in "COLONIALISM ON TRIAL: Indigenous Rights and the Gitksan and Wet'suwet'en Sovereignty Case", Don Monet and Skanu'u (Ardythe Wilson), (Philadelphia/Gabriola Island: New Society Publishers, 1992), x

75 See: Christopher McKee, *TREATY TALKS IN BRITISH COLUMBIA*, Ibid, 30-31

need to develop natural resources can, in principle, justify infringement of Aboriginal title by the Province[76] (raising the duty to consult and accommodate); on the other hand, it means that an Aboriginal group can choose economic uses for its land that are not confined to its culture or historical practices.[77]

The Court then set out three components as the basis for a successful claim of Aboriginal title:

- The land must have been occupied prior to European sovereignty (in British Columbia, that was 1846);
- If the proof of pre-sovereignty occupation to be relied upon is 'current possession', then there must be continuity between pre-sovereignty and present occupation; and
- At the time that European sovereignty was established, Aboriginal occupation must have been exclusive, or with provision for shared exclusivity.[78]

The Court stated that Aboriginal title cannot be surrendered to the Province: where it has not been surrendered to or otherwise extinguished by the federal government, it remains an encumbrance on Provincial Crown title.

Oral History

In a significant clarification of the Canadian law of evidence, Chief Justice Lamer ruled that the trial judge erred in refusing to take into account the Aboriginal oral histories presented to the court by the Gitxsan and Wet'suwet'en to establish use and occupation of their traditional territories, and concluded:

> "The trial judge's treatment of the various kinds of oral histories did not satisfy the principles I laid down in *Van der Peet* ... The trial judge, after refusing to admit, or giving no independent weight to these oral histories, reached the conclusion that the appellants had not demonstrated the requisite degree of occupation for "ownership". Had the trial judge assessed the oral histories correctly, his conclusions on these issues of fact might have been very different."
> (at Para. 107)[79]

Infringement of Aboriginal Rights and the Crown Duty of Consultation

The Court confirmed its ruling in Sparrow[80] that Aboriginal title is not absolute and may be infringed upon by both federal and provincial governments, provided that the two-part Sparrow justification test is met (see page 34).

76 For discussion see: Willms and Kearns, ***The Haida Nation** v. **Weyerhaeuser***, Ibid, 2

77 See: McKee, *TREATY TALKS IN BRITISH COLUMBIA*, Ibid, 89 "The judgment pointed out mining, lumbering, and oil and gas extraction as possible land-related activities that can be pursued by Native peoples on their Aboriginal lands. However, the Court made it clear that this right to choose the uses of the land should be consistent with the Aboriginal group's attachment to the area. For example, if hunting practices were shown to be evidence of attachment to or occupation of land claimed as Aboriginal title land, the relevant Native group could not strip-mine the area."

78 Ibid.

79 The Gitxsan First Nation rightly attach great significance to this oral history ruling, as they argue, "This is an important ruling for future **Delgamuukw** action, and for all future First Nation court cases, in that oral history will now be given as much weight as written evidence." See: www.gitxsan.com/html/delga.html/delgamuukw/oralhistory and see also **Delgamuukw**, Ibid at Para 87: "Notwithstanding the challenges created by the use of oral histories as proof of historical facts, the laws of evidence must be adapted in order that this type of evidence can be accommodated and placed on an equal footing with the types of historical evidence that courts are familiar with, which largely consists of historical documents."

80 R. v. Sparrow, [1990] 1 S.C.R 1075

In elaborating on the implication and weight of that special fiduciary duty of the Crown to look after the best interests of Aboriginal peoples, the Supreme Court also addressed the issue of **consultation**:

> "The fiduciary duty between the Crown and Aboriginal Peoples may be satisfied by the involvement of Aboriginal peoples in decisions taken with respect to their lands. There is always a duty of consultation. Whether the Aboriginal group has been consulted is relevant in determining whether the infringement of Aboriginal title is justified ... The nature and scope of the duty of consultation will vary with the circumstances. In occasional cases, when the breach is less serious or relatively minor, it will be no more than a duty to discuss important decisions that will be taken with respect to lands held pursuant to Aboriginal title ... **The minimum acceptable standard is consultation [that] must be in good faith, and with the intention of substantially addressing the concerns of the Aboriginal peoples whose lands are at issue. In most cases, it will be significantly deeper than mere consultation.** Some cases may even require the full consent of an Aboriginal nation, particularly when provinces enact hunting and fishing regulations in relation to aboriginal lands." [81]

The Court went further to note that the "inescapable economic component" of Aboriginal title referred to earlier means that compensation will be required when that title is infringed. The amount and nature of the compensation required will depend on the context – i.e. on "... the nature of the Aboriginal title affected, the severity of the infringement, and the extent to which Aboriginal interests have been accommodated."[82]

81 Delgamuukw, Ibid, para. 168 For a good analysis of "The Scope of Constitutional Protection of Aboriginal Title", see: McKee, *TREATY TALKS IN BRITISH COLUMBIA*, Ibid, 90-91

82 Delgamuukw, Ibid, para. 169, McKee, Supra, 91

Haida Nation v. British Columbia (Minister of Forests)[83] and *Taku River Tlingit First Nation v. British Columbia*[84], Court of Appeal of British Columbia and Supreme Court of Canada, 2002-2004

On November 18, 2004, the Supreme Court of Canada handed down its eagerly awaited decisions in these companion cases. Both were appeals from the British Columbia Court of Appeal, which had confirmed that the provincial government must properly consult with and accommodate the interests of First Nations before taking or allowing actions that may infringe on Aboriginal title or rights.

In February 2002, the British Columbia Court of Appeal had ruled unanimously that the Crown Provincial and Weyerhaeuser Company Limited did not properly consult the Council of the Haida Nation when renewing a tree farm licence on Haida Gwaii (Queen Charlotte Islands). Tree Farm Licence 39, issued to Weyerhaeuser, contains several areas of old growth red cedar – a culturally significant tree used for totem poles, canoes, and log houses. The Haida Nation wanted large areas of old growth forest protected from clear cutting and its potential detrimental effects on land, watersheds, fish, and wildlife. The appeal court held that "there is a reasonable possibility that the Haida will be able to establish Aboriginal title to at least some parts of the coastal and inland areas of Haida Gwaii as well as an Aboriginal right to harvest red cedar trees from the various old-growth forests on Haida Gwaii."

The duty to consult was found to derive from the 'trustee and beneficiary' relationship between the Crown and First Nations, created when a First Nation asserts title through entering the treaty process, and continuing until after a treaty is signed or Aboriginal rights and title are defined through the courts. The Court of Appeal ruled that the nature and extent of that duty to consult depends on the strength of the First Nation's connection to the land, and concluded that the Haida Nation had a potentially strong legal claim to Aboriginal rights and title.

The Court went on to make the unprecedented ruling that private parties whose activities potentially interfere with or infringe upon Aboriginal rights owe obligations to consult with affected Aboriginal peoples, concluding that:

> "Both the Crown Provincial and Weyerhaeuser had an obligation to consult the Haida people in 1999 and 2000 about accommodating the aboriginal title and aboriginal rights of the Haida people when consider-

[83] Haida Nation v. British Columbia (Minster of Forests), (2002), 5 B.C.L.R. (4th) 33, 2002 BCCA 462, 2004 SCC 73

[84] Taku River Tlingit First Nation v. British Columbia (Project Assessment Director), (2002), 98 B.C.L.R. (3d) 16, 2002 BCCA 59, 2004 SCC 74

> ation was being given to the renewal of Tree Farm Licence 39 and Block 6. The obligation extended to both the cultural and economic interests of the Haida people. Similar obligations had existed much earlier and had continued through until 1999 and 2000 and on to the present time."[85]

The B.C. Court of Appeal's *Haida Nation* ruling followed its January 31, 2002 decision in *Taku River Tlingit*.[86] The Taku River Tlingit First Nation had environmental concerns about the proposed reopening of the Tulsequah Chief Mine in Northwest B.C., which would require the construction of a 160-kilometre access road through one of the largest and most pristine watersheds in the province. They petitioned the B.C. Supreme Court to set aside the Project Approval Certificate (PAC) approving the development. The chambers judge ruled that the provincial ministers who had granted the PAC should have been mindful that their decision might infringe on Aboriginal rights, and that they hadn't been careful enough in the final months of the environmental assessment process to ensure that the substance of the Taku River Tlingit's concerns were addressed. The majority of the Court of Appeal upheld that ruling, finding that the Province had failed to meet its duty to consult with and to accommodate the Taku River Tlingit First Nation.

While the Supreme Court of Canada's decision in *Delgamuukw* confirmed the validity and importance of Aboriginal title and set out strong principles to govern Crown infringement on Aboriginal title, widespread uncertainty was raised by the broad sweep of its language on the nature and various duties of *consultation.*

The B.C. Court of Appeal decisions in *Haida Nation* and *Taku River Tlingit* turned that uncertainty into outright confusion (both for the Crown and for industry), regarding the source of the duty to consult (i.e. the *Sparrow* test for justifying infringement on Aboriginal title and rights, or trust law, or otherwise), and regarding the determination of when that duty arises, what exactly must be done to satisfy it, and who owes the duty. All levels of government, Aboriginal communities, and industry hoped that the Supreme Court would provide clarity and direction in its treatment of the both the *Haida Nation* and *Taku River Tlingit* cases.

85 Willms and Kearns, ***The Haida Nation** v **Weyerhaeuser***, Ibid, 8-9. Note that Charles F. Willms acted as counsel in both these Supreme Court of Canada cases.

86 It is noteworthy that the Taku River Tlingit had participated in the provincial environmental assessment process, and that their petition to set aside the Province's granting of the Environmental Assessment Certificate was brought on grounds of administrative law and on its Aboriginal rights and title - not on the issues of consultation or accommodation.

Haida Nation – Supreme Court of Canada, November 18, 2004

By a unanimous (7-0) decision delivered by Chief Justice McLachlin, the Supreme Court of Canada went a long way toward providing that clarity and direction. The strongly worded judgment makes two issues very clear. First, both orders of government have an inescapable constitutional duty to consult and accommodate Aboriginal communities, in a manner that is meaningful, timely and reflective of the "honour of the Crown", regarding potential infringement on an Aboriginal right or title. Second, that duty rests with the Crown; it cannot be delegated to and does not otherwise extend to third parties (i.e. to industry).

The Crown's Duty to Consult and Accommodate

The Court's thorough and clear explanation of the source, nature, and object of the duties of consultation and accommodation is more than just a stern message to the Government of British Columbia; it also provides valuable guidance to both industry and Aboriginal leadership for successful engagement. Going to the heart of the purpose of this Guidebook, Chief Justice McLachlin's judgment in *Haida Nation* deserves to be reviewed with care here.

Source of the Duty

"The government's duty to consult with Aboriginal peoples and accommodate their interests is grounded in the honour of the Crown. The honour of the Crown is always at stake in its dealing with Aboriginal peoples ... The historical roots of the principle of the honour of the Crown suggest that it must be understood generously in order to reflect the underlying realities from which it stems. In all its dealings with Aboriginal peoples, from the assertion of sovereignty to the resolution of claims and the implementation of treaties, the Crown must act honourably. **Nothing less is required if we are to achieve "the reconciliation of the pre-existence of Aboriginal societies with the sovereignty of the Crown."** (Delgamuukw at para. 186)"[87]

These opening statements from the Court's analysis of the "source of a duty to consult and accommodate" cut to the heart of the matter. Ultimately, that "reconciliation of the pre-existence of Aboriginal societies with the sovereignty of the Crown" has to be found before the legacy of expropriation and assimilation can be put behind us, and before lasting business certainty can be achieved in British Columbia.

[87] Supra, paras. 16-17

When the Duty to Consult Arises

The Province presented its traditional argument that any duty of consultation or accommodation that it might have does not arise unless and until a claim of Aboriginal title or rights is proven. The Court flatly rejected that view:

> "[T]he duty to consult and accommodate is part of the process of fair dealing and reconciliation that begins with the assertion of sovereignty and continues beyond formal claims resolution. Reconciliation is not a final legal remedy in the usual sense. Rather, it is a process flowing from the rights guaranteed by s. 35(1) of the *Constitution Act, 1982*. This process of reconciliation flows from the Crown's assertion of sovereignty over an Aboriginal people and *de facto* control of land and resources that were formally in the control of that people ... To limit reconciliation to the post-proof sphere risks treating reconciliation as a distant legalistic goal, devoid of the "meaningful content" mandated by the "solemn commitment" made by the Crown in recognizing and affirming Aboriginal rights and title ... It also risks unfortunate consequences. When the distant goal of proof is finally reached, the Aboriginal peoples may find their land and resources changed and denuded. This is not reconciliation. Nor is it honourable." [88]

While this process of "reconciliation and honour" may be a fine moral and legal starting point, let's cut to the practical problem:

> **"But, when precisely does a duty to consult arise? The foundation of the duty in the Crown's honour and the goal of reconciliation, suggest that the duty arises when the Crown has knowledge, real or constructive, of the potential existence of the Aboriginal right or title and contemplates conduct that might adversely affect it."**[89]

Scope and Content of the Duty to Consult and Accommodate

If the duty arises before a claim of Aboriginal title or rights is proven, how is the Crown (and industry, as a matter of good business relations) to know what range and depth of consultation (and perhaps accommodation) is appropriate?

> "[I]t will frequently be possible to reach an idea of the asserted rights and of their strength sufficient to trigger an obligation to consult and accommodate, short of final judicial determination or settlement. To facilitate this determination, claimants should outline their claims with clarity, focussing on the scope and nature of the Aboriginal rights they assert and on the alleged infringements ...

[88] Supra, paras. 32-34

[89] Supra, para. 35

There is a distinction between knowledge sufficient to trigger a duty to consult and, if appropriate, accommodate, and the content or scope of the duty in a particular case. The content of the duty, however, varies with the circumstances ... A dubious or peripheral claim may attract a mere duty of notice, while a stronger claim may attract more stringent duties ... **Parties can assess these matters, and if they cannot agree, tribunals and courts can assist. Difficulties associated with the absence of proof and definition of claims are addressed by assigning appropriate content to the duty, not by denying the existence of the duty.**"[90]

In essence, the Supreme Court has ruled that the Crown's duty to consult and accommodate regarding potential infringement on claims of Aboriginal title and rights is **proportionate** – to the strength of the particular claim and to the potential harm that may be caused to that claim by the proposed policy or action.

"Accommodation": Meaning and Mutual Responsibility

The Supreme Court also provided valuable clarification of the working meaning of "accommodation", and set out some pointed advice clearly intended for both government and Aboriginal leaders:

"At all stages, good faith on both sides is required. The common thread on the Crown's part must be "the intention of substantially addressing [Aboriginal] concerns" as they are raised (*Delgamuukw* at para. 168), through a meaningful process of consultation. As for Aboriginal claimants, they must not frustrate the Crown's reasonable good faith attempts, nor should they take unreasonable positions to thwart government from making decisions or acting in cases where, despite meaningful consultation, agreement is not reached ...

The controlling question in all situations is what is required to maintain the honour of the Crown and to effect reconciliation between the Crown and the Aboriginal peoples with respect to the interests at stake. Pending settlement, the Crown is bound by its honour to balance societal and Aboriginal interests in making decisions that may affect Aboriginal claims. The Crown may be required to make decisions in the face of disagreement as to the adequacy of its response to Aboriginal concerns. Balance and compromise will then be necessary."[91]

The issue of "accommodation" flows from the practice of good faith consultation, which in certain circumstances may reveal a duty to accommodate:

"Where a strong *prima facie* [at first sight] case exists for the claim, and the consequences of the govern-

90 Supra, paras. 36-37
91 Supra, paras. 42-46

ment's proposed decision may adversely affect it in a significant way, addressing the Aboriginal concerns that may require taking steps to avoid irreparable harm or to minimize the effects of infringement, pending final resolution of the underlying claim ... Accommodation is achieved through consultation.

The process does not give Aboriginal groups a veto over what can be done with land pending final proof of the claim. The Aboriginal "consent" spoken of in *Delgamuukw* is appropriate only in cases of established rights, and then by no means in every case. Rather, what is required is a process of balancing interests, of give and take."[92]

Who Owes the Duty?

The Supreme Court determined that the Crown must bear sole responsibility for consultation and accommodation regarding the possibility of infringement of Aboriginal title or rights by its actions, or by actions performed under its licence. Rejecting the trust law analysis that had been used by the B.C. Court of Appeal to impose the duty of consultation on Weyerhaeuser, Chief Justice McLachlin focused once again on "sovereignty" and "the honour of the Crown":

"[T]he duty to consult and accommodate ... flows from the Crown's assumption of sovereignty over lands and resources formerly held by the Aboriginal group ... The Crown alone remains legally responsible for the consequences of its actions and interactions with third parties, that affect Aboriginal interests. The Crown may delegate procedural aspects of consultation to industry proponents seeking a particular development; this not infrequently done in environmental assessments. Similarly, the terms of T.F.L. 39 mandated Weyerhaeuser to specify measures that it would take to identify and consult with 'aboriginal people claiming an aboriginal interest in or to the area' (Tree Farm Licence No. 39, Haida Tree Farm Licence, para. 2.09(g)(ii)). **However, the ultimate legal responsibility for consultation and accommodation rests with the Crown. The honour of the Crown cannot be delegated."**[93]

Industry Still Has Legal Responsibility for Its Actions

It is important to note that this decision leaves industry and other third parties fully accountable for their own actions in dealing with Aboriginal title and rights:

"The fact that third parties are under no duty to consult or accommodate Aboriginal concerns does not mean that they can never be accountable to Aboriginal peoples. If they act negligently in circumstances where they owe Aboriginal peoples a duty of care, or if they breach contracts with Aboriginal peoples or deal with them dishonestly, they may be held legally liable."[94]

92 Supra, para. 47-48
93 Supra, para. 53
94 Supra, para 56

While the Court did not spell out the "circumstances where third parties may owe Aboriginal peoples a duty of care", the clarity of this Supreme Court definition of the depth and scope of the Crown's duty to consult and to accommodate will certainly inform third parties' duty of care. 'Foreseeability' of the harm that may be caused by one's actions to someone owed a duty of care goes to the heart of liability in the law of negligence. "Consultation and accommodation" remains prudent business practice in situations where Aboriginal title or rights (whether claimed or confirmed) may be infringed upon by business decisions or actions.

As Anne Giardidi, Weyerhaeuser Assistant General Counsel, observes, "I don't see this as a 'get out of jail free pass' for business at all. When it is your project, your business, your daily activity, you are going to want to ensure that what the Supreme Court of Canada is saying gets done."[95] And, Judith Sayers, Chief Councillor of Hupacasath First Nation, gives industry this piece of good advice:

"Third parties may not have a legal duty to consult First Nations, but it sure makes good business sense. If consultation isn't done properly, a First Nation could go to court for judicial review of the decision and have the permit or license quashed. Can you take that risk? We can enter into a new era of doing business in BC based on good relations, by working with First Nations to protect what is important to them and still doing smart business."[96]

The Province's Duty to Consult, Accommodate, and Negotiate

British Columbia's argument that any duty to consult or accommodate belongs solely to the federal government was dismissed summarily by the Court:

> "[T]he Province took [its] interest in the land subject to "any interest other than that of the Province in the same." The duty to consult and accommodate here at issue is grounded in the assertion of Crown sovereignty which pre-dated the Union. It follows that the Province took the lands subject to this duty. It cannot therefore claim that s. 35 [of the *Constitution Act, 1982*] deprives it of powers it would otherwise have enjoyed."[97]

Throughout the *Haida Nation* judgment are powerful statements linking the Province's duty to consult and accommodate with its parallel duty to negotiate treaties "with honour and integrity, avoiding even the appearance of "sharp dealing"" (*Badger*, at para. 41):[98]

> "The honour of the Crown also infuses the processes of treaty making and treaty interpretation ...Where treaties remain to be negotiated, the honour of the Crown requires negotiations leading to a just settlement of

[95] Anne Giardini, Assistant General Counsel of Weyerhaeuser Company Limited, in Peter Kennedy's "***B.C. companies laud top court's rulings on native consultation***", The Globe and Mail, November 19, 2004, B3

[96] Dr. Judith Sayers, ***Post Haida***, Unpublished Article, November, 2004

[97] *Haida Nation*, Ibid, para. 59

[98] Supra, para. 19

Aboriginal claims … Treaties serve to reconcile pre-existing Aboriginal sovereignty with assumed Crown sovereignty, and to define Aboriginal rights guaranteed by s. 35 of the *Constitution Act, 1982.* Section 35 represents a promise of rights recognition, and "[i]t is always assumed that the Crown intends to fulfill its promises" (Badger, at para. 41). This promise is realized and sovereignty claims reconciled through the process of honourable negotiation …

Put simply, Canada's Aboriginal peoples were here when the Europeans came, and were never conquered. Many bands reconciled their claims with the sovereignty of the Crown through negotiated treaties. Others, notably in British Columbia, have yet to do so. The potential rights embedded in these claims are protected by s. 35 … The honour of the Crown requires that these rights be determined, recognized and respected. This, in turn, requires the Crown, acting honourably, to participate in processes of negotiation … Honourable negotiation implies a duty to consult with Aboriginal claimants and conclude an honourable agreement reflecting the claimants' inherent rights."[99]

Taku River Tlingit – Supreme Court of Canada

In this companion case to *Haida Nation*, the Court's unanimous (7-0) decision was also written by Chief Justice McLachlin. Applying *Haida Nation*, the Court ruled that the Crown owed a duty to consult meaningfully with the Taku River Tlingit First Nation (TRTFN) regarding the decision to re-open the Tulsequah Chief Mine, and to permit Redfern Resources Ltd. (the mine operator) to build a 160 kilometre access road through a portion of their traditional territory. However, the Chief Justice also found that duty of consultation to have been satisfied by the nature and extent of the TRTFN's involvement in the three and a half year review process conducted under B.C.'s Environmental Assessment Act:

> "Within the terms of the process provided for project approval certification under the Act, TRTFN's concerns were adequately accommodated. In addition to the discussion in the minority report, the majority report thoroughly identified the TRTFN's concerns and recommended mitigation strategies, which were adopted into the terms and conditions of certification. These mitigation strategies included further directions to Redfern to develop baseline information, and recommendations regarding future management and closure of the road.

[99] Supra, paras. 19-26

> Project approval certification is simply one stage in the process by which a development moves forward. In _Haida_, the Province argued that although no consultation occurred at all at the disputed, strategic level, opportunities existed for Haida input at a future "operational" level. That can be distinguished from the situation in this case, in which the TRTFN was consulted throughout the certification process and its concerns accommodated."[100]

We hope that our readers recognize the clarity, force and fairness of the Supreme Court's decisions in _Haida Nation_ and _Taku River Tlingit_. We also hope that our review of the line of key Supreme Court rulings on Aboriginal title and rights, going back to _Calder_, has illustrated the legal foundation built by the Court for the Crown's 'triple-decker' duty of consultation, accommodation and negotiation. That foundation is: Aboriginal title and rights claims forming the subject of litigation and treaty negotiations must be respected and protected by both the federal and provincial levels of government, while such litigation and treaty negotiations are pending.[101]

Haida Nation Implications

In closing our discussion of this subject, we note that the Supreme Court's _Haida Nation_ decision presents several important implications for both the mining and Aboriginal communities.

First, the Crown cannot delegate its constitutional duty to consult and accommodate regarding potential infringement of Aboriginal rights and title, whether claimed or proven. Second, the Crown duty to consult and accommodate does not relieve industry of legal responsibility for its actions. Third, all parties should be reasonable, and should be prepared to negotiate. Aboriginal interests should not interpret their right to consultation and accommodation as a veto.

Finally, industry should be wary about relying on government to conduct Aboriginal consultation and accommodation on its behalf.[102] Herb George, Chair of the First Nations Governance Centre, closed a recent presentation at "THE ABORIGINAL ENGAGEMENT & SUSTAINABILITY CONFERENCE - Building Sustainable Relationships" (Vancouver: February 8-9, 2005) with this advice to industry:

> "Don't rely on government to consult for you. **For the same reasons that compelled the Supreme Court to spell out in _Haida Nation_ that government cannot delegate its duties of Aboriginal consultation and accommodation to industry, you should be very reluctant to delegate conduct of your Aboriginal business interests to government.**"

100 _Taku River Tlingit_, Ibid, paras. 44-45

101 For two clear and concise commentaries by leading legal practitioners in the field of Aboriginal rights and title, see: Charles F. Willms and Kevin O'Callaghan, ***The Supreme Court of Canada Decisions in Haida and Taku: The Final Word on the Duty to Consult***, in "Aboriginal Bulletin", Vancouver: Fasken Martineau, November, 2004, www.fasken.com/web/fmdwebsite.nsf/AllDoc/3A6AEB410DC8492185256F5000717B9F/$File/BULLETIN_ABORIGINAL_NOVEMBER_2004.PDF!OpenElement ; and Tom Isaac and Tony Knox, *The Crown's Duty to Consult and Accommodate Aboriginal People: The Supreme Court of Canada's Decisions in Haida Nation v. B.C. and Weyerhaeuser and Taku River Tlingit First Nation v. B.C.*, in "Legal Update", Vancouver: McCarthy Tetrault's Aboriginal Law Group, January, 2005, www.mccarthy.ca
For the perspective of the Haida Nation's legal counsel on the Haida decision, see: www.eaglelaw.org/newsandevents/pr11182004

102 See the comment by Chief Judith Sayers of the Hupacasath First Nation on page 44.

ABORIGINAL POPULATION

Pre-Contact

Until recently, the accepted estimate of the pre-contact Aboriginal population in British Columbia was less than 100,000. The Census of 1871 registered 102,358 Aboriginal people.[103] This low number resulted in part from failure to recognize the extent of early post-contact epidemics, and in part from the scarcity of archaeological research.[104] More recent estimates put B.C.'s pre-contact Aboriginal population between 250,000 and 400,000.

2001 Census

According to the 2001 Canada Census, there were over 170,000 Aboriginal People living in British Columbia. B.C.'s Aboriginal population has been increasing at a greater rate than the rest of the provincial population. Be advised that there are always at least two troublesome issues when working with statistics. The first is how statistics should be interpreted. The second issue revolves around the accuracy of the numbers. There are many different ways of defining and measuring Aboriginal population in Canada. Different agencies have different numbers. According to the 2001 Canada Census[105], 976,305 of 29,639,035

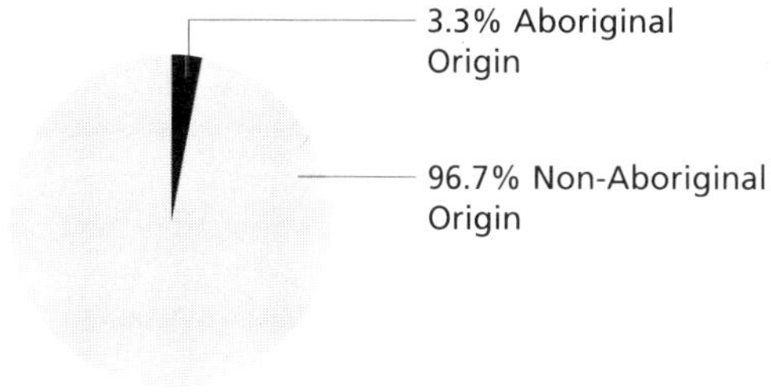

Canadians reported Aboriginal origins. Aboriginal People[106] account for 3.3 per cent of Canada's total population.

First Nations in British Columbia and Canada

The distribution of Canada's First Nation population by province and territory shows that British Columbia First Nations make up 20.4 per cent of Canada's Status Indians. These are Aboriginal People who are registered as "Indians" and have rights and benefits as specified under the *Indian Act*.[107]

Where do Aboriginal People reside?

In the 2001 Canada Census, 49 per cent of the population who identified themselves as Aboriginal People lived in urban areas. It should also be noted that 68 per cent of the Métis population lived in urban areas. In British Columbia, Status Indians are the main group of Aboriginal People who have a land base defined in the *Indian Act* as a "reserve".[108]

103 Source: Statistics Canada, www.statcan.ca

104 Paul Tennant, ***Aboriginal Peoples and Politics***, (Vancouver: UBC Press, 1990)

105 Statistical information can be found at www.statcan.ca .

106 "Aboriginal People" is a broad term, generally used inclusively in Canada to embrace Status Indians, Non-Status Indians, Métis and Inuit.

107 Indian as defined by the ***Indian Act***, Ibid - means a person who pursuant to this Act is registered as an Indian or is entitled to be registered as an Indian.

108 Reserve as defined by the ***Indian Act***, Ibid. - (a) means a tract of land, the legal title to which is vested in Her Majesty, that has been set apart by Her Majesty for the use and benefit of a band, and (b) except in subsection 18(2), sections 20 to 25, 28, 36 to 38, 42, 44, 46, 48 to 51, 58 to 60 and the regulations made under any of those provisions, includes designated lands.

STATUS INDIANS IN B.C. AND REST OF CANADA

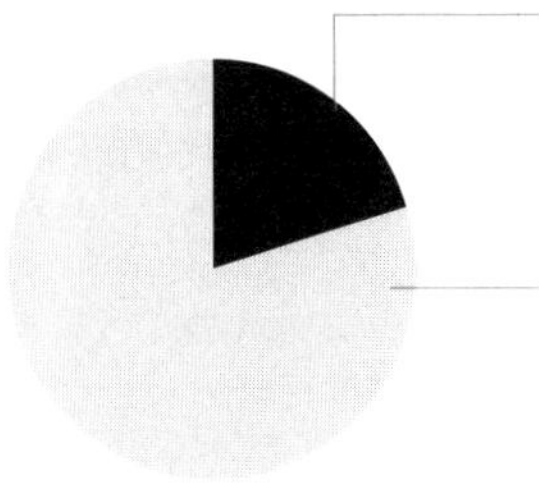

British Columbia's Status and Non-Status Indians make up about 4.1 per cent of the total provincial population (170,025 of 4,141,272, in the 2001 Census).

About one-fifth of British Columbia's Aboriginal population is made up of Non-Status Indians. These are people who are not registered under the *Indian Act*. They are not members of bands and are not entitled to the rights and benefits specified in the *Indian Act*. Some Non-Status Indians had their status reinstated with the passage of Bill C-31 (an amendment to the *Indian Act*) in 1985.

BRITISH COLUMBIA'S ABORIGINAL POPULATION IN RELATION TO TOTAL PROVINCIAL POPULATION

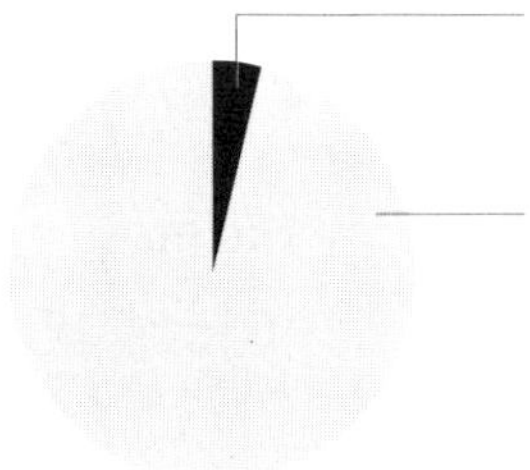

British Columbia Bands

The 197 British Columbia First Nation Bands make up about 33 per cent of the 605 bands in Canada.

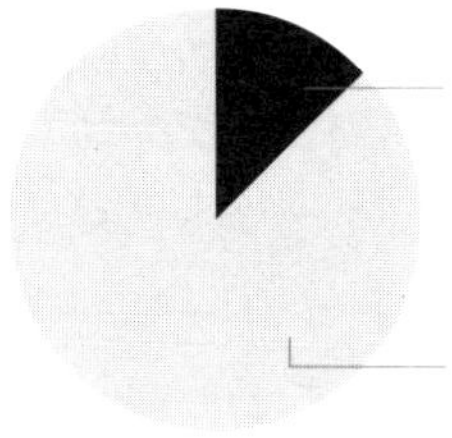

A "band" is a group of people that holds reserve land, has funds held for it by the federal government, or has been declared a band by the Governor-in-Council.[109] A band may hold reserve lands with or without habitation for traditional use. O'Neill points out that confusion can arise from the often interchangeable use of the terms "band" and "First Nation."[110]

Reserves

Number of British Columbia Reserves

First Nations have 1,613 reserves in British Columbia. Reserves were initially created by colonial Governors and later by the Canadian government. They are parcels of land allocated to bands and held in trust by the federal Crown for Status Indians. It is noteworthy that 71 percent of Canada's 2,263 reserves are located in British Columbia.

109 *Indian Act*, Ibid, at section 2(1).

110 J. Stephen O'Neill, *DECISION MAKING ON RESERVES - THE CURRENT SITUATON*, in "Aboriginal Issues Today: A Legal and Business Guide", Ibid, 98: "There is no definition of ***First Nation*** in the ***Indian Act***, but the term has become relatively popular among a good number of Canada's 608 Indian Bands. (It is important to note, however, that while Indian bands are often referred to as First Nations, all First Nations may not necessarily be Indian bands: Inuit communities, for example, are also First Nations."

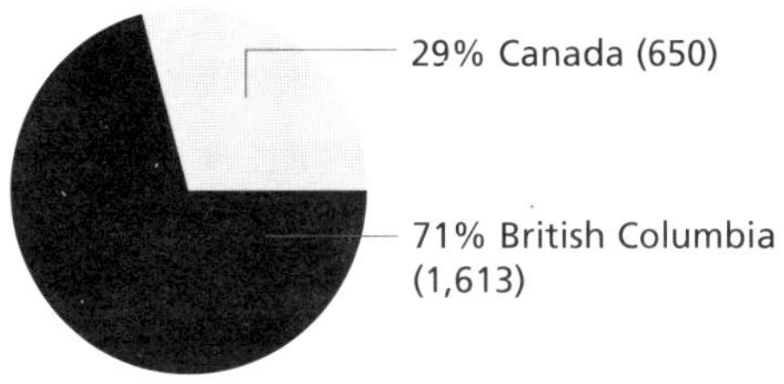

Area of British Columbia Reserves

It is also noteworthy that the area occupied by First Nation Reserves in British Columbia (343,741 hectares) accounts for only 13 per cent of the total area of Canadian reserves (2,684,448 hectares).

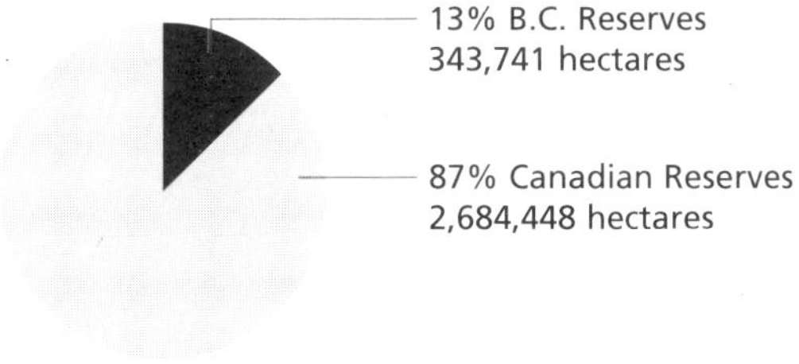

British Columbia Reserve Land

British Columbia's First Nation Reserves account for only 0.36 per cent of the total provincial land area.

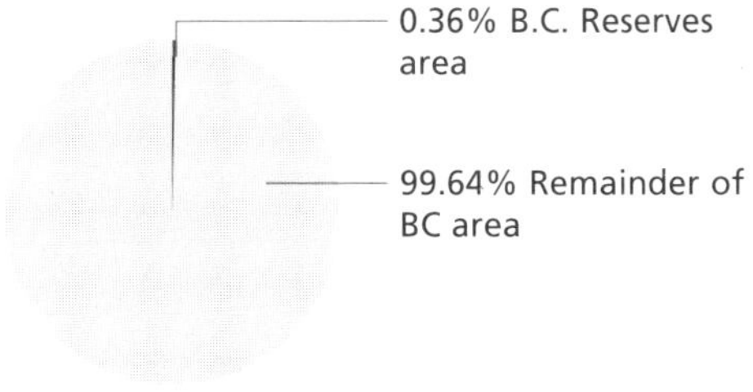

ABORIGINAL PEOPLE: YESTERDAY AND TODAY

Linguistic Diversity

Within British Columbia lie parts of three of the ten commonly recognized North American Aboriginal "culture areas" that flourished from the sixteenth through the early nineteenth century – the Northwest Coast, the Interior Plateau, and the Subarctic.[111] The Northwest Coast was the most densely populated area, followed by the Interior Plateau and then the Subarctic.

Clearly defined geographic boundaries and the abundance of fish, wildlife and natural resources allowed many of B.C.'s Aboriginal communities to develop distinct, thriving cultures and languages. It is believed that as many as fifty distinct Aboriginal languages were being spoken in British Columbia at the time of European contact.[112] Each First Nation had its own language, culture, social structure, legal system and political system.

The diversity and distinctness of First Nations can be seen in the number of Aboriginal language families that exist throughout Canada. Linguists refer to groups of languages that are clearly distinct, yet share enough "cognate vocabulary" to suggest common ancestry and origin as "language families".

111 Robert J. Muckle, *THE FIRST NATIONS OF BRITISH COLUMBIA*, Ibid, 33
112 Supra, 35

Canada is home to eleven distinct Aboriginal language families, of which seven exist in British Columbia.[113] Sixty four percent of Canada's unique Aboriginal languages exist only in British Columbia.

Each language family is completely distinct; the specific languages within a family "... may be as similar as Spanish is to French, and languages belonging to different language families may be as different as English is from Cantonese."[114] If someone from the Athapascan tried to speak to someone from the Wakashan, they would share little ability to communicate. Canada's First Nation communities are made up of people from many different cultures and languages.

There is no such thing as generic or homogenous "Aboriginal People of Canada." That would be like referring to all people from Europe as Europeans. When a Scot is asked "Where did your ancestors come from?" he will answer, "Scotland." If the questioner responds, "Oh, you're European" – the Scot may well be offended. This is no different in Aboriginal communities. If the response to "Haida Gwaii" is "Oh, you're an Indian" – you have likely offended the person of Haida ancestry.

Governance Structures: Customary and Statutory, Hereditary and Elected

There were many different governing institutions in Canada's pre-contact Aboriginal societies. Scholars have transcribed oral histories of matrilineal, patrilineal, and band equalitarian governments. In matrilineal communities women held the balance of power and in patrilineal communities men held the balance of power.

In the 1880's, Section 74 was inserted into the *Indian Act*, imposing a regime for the election of band councils under a system of rules patterned after municipal law. This imposition of a foreign leadership and governance structure has been strenuously opposed over the years by many First Nations, and has contributed to breakdowns of traditional culture and governance in many communities where it has been imposed.[115] Over the years, many bands have struggled to maintain their customary and often hereditary governance culture. More recently, many bands have taken advantage of *Indian Act* amendments allowing a return to customary and hereditary governance structures and systems.[116] While some First Nations have formal clan structures, others organize around elaborate "house" structures. Still others have completely different forms of social organization.

113 See Muckle, Ibid, 35 for a table setting out B.C.'s language families and member languages. Muckle lists eight language families: Algonquian, Athapascan, Haida, Ktunaxa, Salishan, Tsimshian, Tlingit and Wakashan.

114 Supra

115 See: Christopher McKee, "*TREATY TALKS IN BRITISH COLUMBIA*", Ibid, 81-82

116 In 1997, about 50 per cent of the 608 bands registered with the Department of Indian and Northern Development (DIAND) were elected under s. 74 of the *Act*. See O'Neill, Ibid, 108

It is important to understand that pre-contact Aboriginal governance structures often included Hereditary Chiefs (whose powers were passed down from one generation to the next along blood lines or other cultural protocols), and that those structures are legally recognized today. Before commencing negotiations and before relying on band council decisions (whether elected or customary), companies should prepare by learning the particular bands' leadership and decision-making protocols, and then should protect themselves by confirming that the prescribed decision-making procedures were followed.[117] Also, in light of the *Indian Act's* limitations on a band council's ability to contract, it is important to consider legal enforcement mechanisms prior to contract execution.

Aboriginal Youth

As stated above, Aboriginal People accounted for 3.3 per cent of Canada's total population in 2001. Approximately 33 per cent of Aboriginal People were aged 14 and under in 2001, compared to 19 per cent of the same age group in the non-Aboriginal population.

Median Age

The median age is the point where exactly one-half of the population is older and the other half is younger. In the 2001 Census, the median age for non-Aboriginal Canadians was 37.7 years, while for Aboriginal People the median age was 24.7 years.

Fastest Growing Population

The birth rate of Aboriginal children is about 1.5 times higher than that of non-Aboriginal children. Aboriginal children represent 5.6 per cent of all children in Canada.

Forces Affecting Canada's Aboriginal People

In comparison to non-Aboriginal People, Canada's Aboriginal People experience:

- Poorer health,
- Lower levels of education,
- Lower income levels,
- Higher rates of unemployment,
- Higher levels of incarceration, and
- Higher rates of suicide.

Community Needs Drive First Nations' Negotiation Priorities

These issues preoccupy leadership thinking in Aboriginal Canada, and they are often placed high on Aboriginal negotiating agendas. Aboriginal leaders and negotiators will tend to place premium value on measures that increase levels of health, education, and income in their communities, and on measures that decrease levels of unemployment, incarceration, substance abuse and suicide.

117 Supra, 107

Aboriginal Land Claims and Overlapping Traditional Territories: A Prejudicial Misconception

It is a common and prejudicial misconception that British Columbia's First Nations are claiming 110 per cent of the provincial land base. In preparation for negotiations, the British Columbia Treaty Commission process requires First Nations to identify areas in which they have an interest. [118] Keep in mind that First Nations' traditional 'community' perspective on land is very different from the European 'fee simple' view. In fact, it is not uncommon for First Nations to share the use of their lands, and to have overlapping territorial boundaries.

It is also important to recognize that under the British Columbia Treaty Commission process, First Nations are required to work out overlapping boundaries between themselves during their negotiations with Canada and the Province.[119] Finally, it should be noted that the province of British Columbia has negotiation mandates in place that will see only a small portion of land transferred as part of the treaty process, and that will ensure that privately owned lands will not be subject to treaty-based transfer.[120]

The Treaty-Making Process in British Columbia[121]

Introduction

The early and middle history of British Columbia's approach to resolution of claims of Aboriginal title and rights and to treaty-making has already been reviewed (please see pages 19-21). B.C.'s modern era of treaty-making began in the fall of 1990, with then-Premier Vander Zalm's announcement that his government was prepared to open negotiations with B.C.'s Aboriginal leaders (without acknowledging the existence of unextinguished Aboriginal title). On December 3, 1990, the British Columbia Claims Task Force was established. Made up of representatives of the federal and provincial governments and the First Nations Congress (later renamed the First Nations Summit – (FNS)), the Task Force presented its formal report on June 28, 1991, outlining 19 recommendations for treaty negotiation principles and procedures.

On September 21, 1992, representatives of the two orders of government and FNS signed the British Columbia Treaty Commission Agreement, endorsing all 19 Task Force recommendations – which included the formation of the British Columbia Treaty Commission and the establishment of a six stage treaty negotiation process.

118 For an overview map of Treaty Negotiations in British Columbia, see: http://srmwww.gov.bc.ca/dss/initiatives/treaty/Images/Prov_2003/BCSOI_Asize_Oct2003.pdf

119 See: www.bctreaty.net/files_3/sixstages-14.html

120 See B.C. Treaty Commission Annual Report 2004, Ibid, 13 for discussion of various challenges to be met in treaty-making in urban areas where available Crown land is scarce: "Where Crown land is limited, available-for-purchase private land and other elements of the treaty will be critical to settlements in urban areas."

121 This section of our Guidebook relies on Christopher McKee's ***TREATY TALKS IN BRITISH COLUMBIA: Negotiating a Mutually Beneficial Future***, Ibid, for both structure and content. For an excellent overview and analysis of B.C.'s treaty-making history and current process, please refer to Chapters 1 and 2.

The British Columbia Treaty Commission (the Commission)

Appointed on April 15, 1993 as the "keeper of the process" – the Commission is an independent and impartial body, made up of five commissioners: one appointed by each order of government, two by the FNS, and a chief commissioner agreed to by all three principals. The Commission has three complementary roles:

- *Facilitation* – supporting the negotiation process;
- *Funding* – allocating support funding to level the playing field for First Nations' negotiations with the federal and provincial governments; and
- *Public information* – providing non-partisan information on the treaty process to the public, using a variety of communications tools.

The Commission has authority to make recommendations, but is unable to compel parties or principals to act. An annual report on the effectiveness of the process and on the progress of each negotiation is prepared by the Commission for both orders of government, First Nations, and the public.[122]

The Treaty-making Process

Treaty-making is entirely voluntary, and follows the six stage process recommended by the Task Force, which is summarized below:

Stage 1: Statement of Intent

The filing of a Statement of Intent (SOI) by a First Nation starts the treaty process. To be complete, the SOI must identify the First Nation and the people whom it represents, describe the area of the traditional territory, and identify a formal contact person.

Stage 2: Preparations for Negotiations

Within 45 days of receiving the SOI, the Commission must convene a meeting of the three parties to exchange information, review the criteria that will determine readiness to negotiate a treaty, consider the research that may be needed to prepare for negotiation, and outline the main issues of concern to each party. Once the Commission is satisfied that a set of specific criteria has been established, it will confirm that the parties are ready to begin Stage Three.

122 Christopher McKee, Supra, 33. See also: BC TREATY COMMISSION, *Commissioners Adam and Harcourt's Presentation to Supreme Court Justices*, Thursday, November 10, 2004, page 1. (www.bctreaty.net/files_3/annuals.html)

Stage 3: Negotiation of a Framework Agreement

The Framework Agreement is an agenda approved by the three parties that identifies: the issues on the table, the goals of the negotiation process, special procedural arrangements, and the negotiation timetable.

Stage 4: Negotiation of an Agreement in Principle

This stage marks entry into substantive negotiations directed at achieving a series of agreements that will form the basis of a treaty. The Agreement in Principle also lays out the ratification procedure that each party will follow in seeking a mandate from their respective constituents to finalize their treaty.

Stage 5: Negotiation to Finalize a Treaty

The accords reached in the Agreement in Principle will be incorporated into the formal treaty at Stage 5. Remaining technical and legal issues will be addressed, and if they are resolved, a Final Agreement will be signed and formally ratified to complete the negotiation process.

Stage 6: Implementation of the Treaty

Plans and protocol for the long-term implementation of the treaty will be made during this stage of the process.

Funding the Treaty-Making Process

A Memorandum of Understanding (MOU) between the Governments of Canada and British Columbia dated June 21, 1993 established the funding regime for all stages of the treaty-making process. The MOU deals with four broad issues: pre-treaty costs, settlement costs, implementation costs, and self-government costs.[123]

First Nations Participation in the Treaty Process

There are currently 55 First Nations[124] participating in the B.C. treaty process at 44 negotiation tables. Five negotiations are at Stage 5, 41 are at Stage 4, three at Stage 3, and six are at Stage 2. Late in 2004, four First Nations were "cautiously optimistic" that they would "conclude final agreements early in 2005, if they were able to maintain the current pace of intensive negotiations and reach agreement on all of the outstanding issues."[125] Another dozen First Nations hope to achieve Agreements in Principle in 2005 or 2006.

123 The Commission has criticized the B.C. government for concentrating its resources on a handful of negotiation tables, and has urged the Province "... to provide adequate resources to treaty tables where First Nations are ready willing and able to negotiate." BC Treaty Commission Presentation to Supreme Court Justices, Ibid, p. 2. However, in September 2004 the Commission announced, "We were pleased by Minister Plant's announcement last week of an additional $2,135,000 to work toward final agreements and additional agreements in principle. And there is indication from the minister that if we do have success in achieving agreements, more resources could be available next year. See: www.bctreaty.net , BC Treaty Commission Presentation to First Nations Summit, Thursday, September 23, 2004, North Vancouver, 4
McKee points out interesting political and policy implications arising from the terms of the MOU: "First, off-loading between the two orders of government is [prohibited] ... This provision in the MOU should prevent much of the off-loading that the federal government has undertaken in recent years in policy areas that are subject to shared jurisdiction with the provinces. It should also serve as a check against the provincial government assuming a preponderance of the policy responsibilities over Aboriginal affairs once the treaties have been ratified. Apart from where there is devolution of power to First Nations as a result of the treaties, the policy authority will be shared by the federal and British Columbia governments." Christopher McKee, *TREATY TALKS IN BRITISH COLUMBIA*, Ibid, 113

124 Representing about two-thirds of the Aboriginal people of British Columbia. See: BC Treaty Commission Presentation to Supreme Court Justices, Ibid, 2

125 BC Treaty Commission Annual Report, 2004, Ibid, 10

Interim Measures Agreements (IMAs)

The creation of a series of IMAs was one of the British Columbia Task Force recommendations that led to the current treaty process. IMAs are intended primarily to regulate the management and use of natural resources and lands – protecting and balancing the interests of the parties (including claimed Aboriginal title and rights) during the treaty negotiation process.

Two general types of IMAs exist. First, 'program-related interim measures' are used to carry out the responsibilities of various government ministries, including fulfilling the provincial government's fiduciary responsibilities to Aboriginal people. Such IMAs are seen by the Province as integral to its overall responsibility as a government. Second, 'treaty-related measures' are used to protect agreements made during the negotiation process while other issues continue to be negotiated. Treaty-related measure IMAs are intended to safeguard the integrity of negotiation process from the various political and business pressures influencing all of the parties during the long negotiation process.[126]

The B.C. government has sought to demonstrate its commitment to the current treaty process by pointing out that over 300 IMAs have been put in place to date, "... including 69 treaty related measures, 145 economic development projects and 15 agreements for co-management of parklands."[127]

Fair Negotiation of Fair Treaties: the Business Case

Our focus here is to review briefly the economic implications of proceeding with good faith negotiation of fair treaties, or in the alternative, staying with the status quo.

Costs and Benefits of Settling Treaties

There is no doubt that the costs will be considerable. However, there is strong research available that indicates that the benefits will far outweigh those costs. In 2004, the Commission released a comprehensive study prepared by the accounting firm of Grant Thorton LLP, *An Update to the Financial and Economic Analysis of Treaty Settlements in British Columbia*.[128] Included were summaries of two major studies of the economic implications of settling treaties in British Columbia, along with an earlier study of the costs of uncertainty caused by widespread unresolved Aboriginal land claims.[129] In 1996, the accounting firm of KPMG released a study containing the following key findings:

126 Treaty-related IMAs must be submitted to the provincial cabinet for approval. See McKee, Ibid, 41

127 BC Treaty Commission Presentation to FNS, September 23, 2004, Ibid, 4

128 Grant Thorton LLP, *An Update to the Financial and Economic Analysis of Treaty Settlements in British Columbia*, submitted to British Columbia Treaty Commission, March 12, 2004.

129 Supra, Executive Summary, 3-4

- First Nations would accrue a financial benefit of $6.02 billion to $6.63 billion (1995$, over 40 years)
- Non-First Nations British Columbians would incur a financial cost of between $1.37 billion to $2.11 billion (1995$, over 40 years)
- **Net financial benefits to British Columbians were estimated to be $3.91 billion to $5.26 billion (1995$, over 40 years)**[130]

In 1999, Grant Thorton LLP issued its *Financial and Economic Analysis of Treaty Settlements in BC*, including the following key findings:

- First Nations would accrue a financial benefit of $6.28 billion to $6.76 billion (1998$, over 40 years)
- Non-First Nation British Columbians would incur a financial cost between $2.08 billion to $2.48 billion (1998$, over 40 years)
- **Net financial benefits to British Columbians were estimated to be $3.80 billion to $4.68 billion (1998$, over 40 years)**[131]

Finally, we should note an interesting 1995 study commissioned by the governments of Canada and British Columbia. The ARA Consulting Group assessed the overall impact of treaty settlement of Aboriginal claims in B.C., based on the treaty experiences of other parts of Canada, the United States, Australia, and New Zealand. That study's conclusions included the following findings:

- "The reaction of the non-Aboriginal community to settlement has generally been positive.
- In each of the cases Aboriginal mechanisms exist or are planned to carry out business investment.
- Aboriginal groups have taken a variety of approaches to business investment, including investment in a number of non-traditional businesses.
- Aboriginal economic corporations have tended to reinvest any profits that have been generated rather than distribute funds to individuals.
- A range of employment opportunities has existed for individuals external to the settlement region or group, largely because some settlement groups lacked these resources internally ... People involved with business activities and resource development have been employed across a full spectrum of jobs.
- The review of land claims has revealed several general trends:
  - Settlements have not brought about any dramatic changes for the non-Aboriginal community;
  - Controversy prior to and during the negotiation of the claim reduces during the implementation period; and
  - In many cases settlement has led to a stronger partnership between Aboriginal and non-Aboriginal societies.

A fundamental principle for the treaty process is to establish a contract of understanding for the future between the Aboriginal and non-Aboriginal communities. **The review of cases**

130 Summarized in *An Update to the Financial and Economic Analysis of Treaty Settlements in British Columbia*, Supra, 3

131 Summarized in *An Update to the Financial and Economic Analysis of Treaty Settlements in British Columbia*, Supra, 4

demonstrates that this can be developed through shared management of resources, increased commercial interaction in the business community through both private Aboriginal investment and joint venture arrangements, and increased understanding through communication during the negotiation process. An approach toward partnership that provides identity, initiates respect, reduces dependency, and increases certainty is a productive outcome of the settlement process."[132]

Costs of Not Settling Treaties

At least as important as the direct costs of completing treaties, are the costs of the alternative – not completing treaties. The key findings of Price Waterhouse's 1990 study, *Economic Value of Uncertainty Associated with Native Claims in B.C.* are summarized below:

- "$1 billion impact and 1,500 jobs affected
- $50 million per year in lost capital expenditures in the mining industry.
- $75 million per year of expenditures in mining delayed 3 years.
- 100 jobs per year were lost as a result of unsettled land claims.
- A 1% investment premium is needed to compensate for the uncertainty."[133]

The impact of unresolved Aboriginal land claims on investment activity and confidence in British Columbia was also examined in a survey conducted in 2004. One hundred forty three interviews were conducted through a survey of 550 contacts, focusing on businesses with B.C. operations generating annual revenue of at least $10 million, in the following sectors: Forestry, Investment/ Financial, Mining, Oil and Gas, Transportation and Utilities.

The results are summarized below:

- "**Q.** Assuming the BC economy performs as you expect over the next five years, how important will each of the following factors be in your company's investment decisions in BC?

132 ARA Consulting Group Inc., *Social and Economic Impacts of Aboriginal Land Claim Settlements: A Case Study Analysis*, Prepared for Ministry of Aboriginal Affairs, Province of British Columbia, and Federal Treaty Negotiation Office, Government of Canada, (Victoria: Queen's Printer for British Columbia, 1996) 24-35

133 Supra, page 3

A. 67 per cent of survey respondents cited 'unresolved First Nation land claims negotiations' as important in investment decisions – very important (31%) and somewhat important (36%).

- **Q.** Have the unresolved First Nation land claims in BC reduced your company's investment in BC over the past five years, or had no real impact on investment decisions?
  A. One in five companies responding to the survey state that "unresolved First Nations land claims" reduced their investment in BC over the past five years. Six per cent reported a significant reduction in investment.

- **Q.** If a significant number of First Nations land claims were settled, would your company increase its investment over and above what is already planned for the next five years, or would this have no real impact on investment plans?
  A. One in four companies responding to the survey would increase investment if a significant number of First Nation claims were settled. Eight per cent would increase investment significantly." [134]

If merely having the dark cloud of unresolved claims of Aboriginal title and other rights hanging in the air has this kind of negative impact on business and investment confidence, what will happen if impatient and frustrated Aboriginal activists resort to blockades and other direct measures? Aboriginal blockades have been used very effectively in the past to captivate media and public attention, creating significant market turmoil and political pressure. McKee provides an excellent review of this issue:

> "Native blockades ... point to a general feeling of discontent by Aboriginal peoples over the manner in which Canadian governments have traditionally addressed Native grievances ... [B]lockades and other crisis situations seem to be a most effective means of forcing governments to take Native claims seriously, and they suggest, by extension, that Aboriginal peoples are no longer willing to accept the stereotype of themselves as victims in a society dominated largely by non-Aboriginal interests ... To what extent these acts of civil disobedience will continue in the future, perhaps even after the treaties are concluded, is uncertain. **But if blockades are to be used less often by Aboriginal groups, and if they are to be seen as illegitimate expressions of Aboriginal discontent, then the treaties must be concluded expeditiously, and must be seen to be fair, especially by Native groups."** [135]

As we leave this area, let's consider the high social cost of the status quo – for Aboriginal and non-Aboriginal communities alike. In the words of Chief Joseph Gosnell Sr., President of the Nisga'a Tribal Council:

[134] The Mustel Group, *Impact of Unresolved First Nations Land Claims on Investment in B.C.*, February, 2004, www.bctreaty.net/files_3/issues_financial.html#studies
[135] McKee, *TREATY TALKS IN BRITISH COLUMBIA*.,Ibid, 116-117

"The highest cost of all which our opponents fail to factor into the hard math of land claims is the costs to taxpayers of keeping the current system. No one, not us, not non-Natives, approves of the billions of dollars spent annually to keep Aboriginal people in the binding poverty of our tiny reserves, beggars on our own land, sharing no part of its resource wealth."[136]

Reflections on the B.C. Treaty Process

The participation of 55 First Nations at 44 separate tables represents the largest treaty negotiation process ever undertaken in Canada. While progress has been made, there have been many delays and setbacks in the treaty process since its inception in 1993. Too few First Nations are at advanced stages of negotiations, and dark clouds of litigation hang heavy in the air. Also, one-third of B.C.'s Aboriginal peoples are not yet participating in treaty talks, including the membership of the Union of British Columbia Indian Chiefs (UBCIC) (representing many B.C. interior First Nations).

Non-Aboriginal criticism of the treaty process has focused on the involvement of third parties and the public in the talks, and on treaty costs. For example, members of the Treaty Negotiation Advisory Committee (appointed pursuant to a Claims Task Force recommendation in 1992 to represent the interests of B.C.'s major business, labour, and wildlife groups in the talks) have complained that their role is too "peripheral" to the negotiations. As McKee points out, while interest groups and the public should have more substantive involvement during the Framework Agreement and Agreement in Principle stages, great care should be taken not to undermine the integrity of actual negotiations between the participants on the basis of prejudice and bias.[137]

The recent Supreme Court of Canada decisions in *Haida Nation* and *Taku Tlingit* create fresh opportunity and heightened urgency for the achievement of significant treaty negotiation breakthroughs. The Supreme Court has raised Aboriginal expectations significantly – for improved quality of consultation and accommodation regarding potential infringement on their Aboriginal title and rights, and for strengthened government commitment of resources and priority to the treaty-making process.

The political and economic stakes are at least as high as the expectations. Let's hope that fears raised by the seriousness of what's at stake are not allowed to drive negative and defensive decision-making.

136 See McKee, Supra, 65
137 McKee, Ibid, 114-15

Partnership: Shared Resources and Shared Success

Resource Expo 2004

A recent conference (November, 2004) in Vancouver brought together decision-makers from around the world to develop opportunities for Aboriginal communities and their corporate partners. Prominent at *Resource Expo 2004* were the National Aboriginal Business Association (NABA), the Native Investment Trade Association (NITA), and the Indian Resource Council of Canada (IRCA) – all Aboriginal organizations focused on bringing business opportunities and Aboriginal people together. According to Dave Tuccaro, head of NABA,

> "In Canada's northern diamond mines, Aboriginals are doing hundreds of millions of dollars in business, and in the oil sands, Aboriginal businesses did over $400 million in 2003 alone. This is the result of good partnerships that are providing Aboriginals with job and educational opportunities – the means to develop a sustainable future. It's time for more Aboriginals to get busy, the sooner the better."[138]

Another featured *Resource Expo 2004* participant was the ATCO Group, a recognized leader in development of innovative and profitable joint venture companies with Aboriginal communities. Michael Shaw, managing director of Global Enterprises for the ATCO Group,

138 Randall Anthony Mang, ***Aboriginal Resource Development Key to National Prosperity***, The Globe and Mail, November 8, 2004, B8

believes that the lessons ATCO has learned over four decades are keys to building a more socially progressive and profitable future for everyone:

> "ATCO brings expertise and capital, but our partners bring skills we don't have. It makes for better companies. Continuously listening to and consulting with elders, along with trust, respect and integrity, are core to working together. Today we have numerous joint ventures with Aboriginal and Inuit communities where we are equal partners, or as in most cases now, we are the minority partner."

PART II:

Mining & Aboriginal Community Enegagement

CASE STUDIES, POINTERS, PRACTICAL ADVICE & BEST PRACTICES

INTRODUCTION

Mining's Economic Impacts

Canada's mineral wealth is well-known throughout the country and the world. Perhaps not so well-known, to industry insiders immersed in operational detail, or to industry observers focused on footprint issues, is the sheer scale of mining's contribution to the Canadian economy. For more than a century, mining has been one of Canada's largest industries, contributing an average of four per cent of the country's real gross domestic product (GDP) over the past 20 years. In 2003, that contribution amounted to about CAN$41 billion in GDP; with another CAN$4.6 billion added by related sectors providing mining-related equipment and services.[139]

The Mining Industry's Brief to the 61st Annual Mines Ministers' Conference, *Building on our Strengths: A World Leader in Mining*, illustrates the scale of mining companies' contribution to Canada's national and regional economies, balance of trade, and transportation infrastructure:

"[M]ining companies created 389,000 direct jobs (one in every 41 Canadian jobs) [in 2003], paying the highest wages of any industrial sector, with total annual average salaries of $19 billion (excluding benefits). Thousands of additional indirect jobs are also created in rural, urban and Aboriginal communities. In 2003, the industry spent $641 million on exploration, $324 million on research and development, was responsible for 30 per cent of all shares traded on the TSX, and supported more than 2,200 small and medium-sized businesses, including Aboriginal businesses...

Mining operations are not only economically important, but inextricably linked to the Canadian economy. Today, we export 74 per cent of mineral production, representing 13 per cent of total Canadian exports (one in every eight export dollars). Further, mining is responsible for 73 per cent of port volume and 61 per cent of rail freight revenue. As a trading nation, mining is not only an engine of the national economy, but a critical component in ensuring the competitiveness of all Canadian exporters."[140]

139 The Mining Association of Canada (MAC) and the Canadian Mineral Industry Federation (CMIF), *Building on our Strengths: A World Leader in Mining*, A Brief to the 61st Annual Mines Ministers' Conference (Iqaluit, Nunavut, July 20, 2004) 4

140 Supra, 4 to p. 49

Mining's Aboriginal Engagement Progress: The Whitehorse Mining Initiative (WMI)

The very fact that this Guidebook is being produced signifies at least two things: progress is being made in advancing the cause and quality of Aboriginal engagement in Canada's mineral exploration and mining sectors; and recognition that further progress needs to be made. Significant steps in that progress have been achieved through the WMI:

> "... a multistakeholder process, involving the mining industry, federal and provincial governments, labour unions, Aboriginal peoples and the environmental community, that took place from 1992 to 1994.
>
> The Leadership Council Accord, which resulted from the WMI process, adopts a strategic vision for a healthy mining industry in the context of maintaining healthy and diverse ecosystems in Canada, and for sharing opportunities with Aboriginal peoples. It calls for improving the investment climate for investors; streamlining and harmonizing regulatory and tax regimes; ensuring the participation of Aboriginal peoples in all aspects of mining; adopting sound environmental practices; establishing an ecologically based system of protected areas; providing workers with healthy and safe environments and a continued high standard of living; recognition and respect for Aboriginal treaty rights; settling of Aboriginal land claims; guaranteeing stakeholder participation where the public interest is affected; and creating a climate for innovative and effective responses to change."[141]

A key element of that WMI Leadership Council Accord addressed the issue of Aboriginal involvement in the mining industry:

> "Many Aboriginal communities want to become more involved in all aspects of development, including the mineral industry. The mining industry could facilitate socio-economic growth in Aboriginal communities.
>
> **Our Principle**
> *Aboriginal peoples are entitled to opportunities to participate fully in mineral development at all stages of mining and associated industries and at all employment levels.*
>
> **Our Goals**
> *To remove the barriers – real and artificial – to education, workplace, and business opportunities that often prevent Aboriginal peoples from maximizing benefits from the mining industry.*
>
> *To allow increased participation of Aboriginal peoples in all parts of the mining industry, including direct employment and related economic or business opportunities.*
>
> *To allow the mining industry, Aboriginal peoples, and other interested stakeholders to develop formalized*

[141] The Mining Association of Canada, *ABORIGINAL ECONOMIC DEVELOPMENT AND THE CANADIAN MINING INDUSTRY, Presentation to the Standing Committee on Aboriginal Affairs and Northern Development, June 10, 1998*, 1 See: www.mining.ca/english/publications/native.html

partner relationships in which there is a better awareness of respective issues, needs, and concerns, and a higher level of mutual understanding.

To support policies, legislation, and agreements that encourage growth in business relationships between the mining industry and Aboriginal communities.

To ensure regular and open consultations between exploration companies and mine developers, and Aboriginal communities, and to ensure that the Aboriginal communities are involved in decision-making processes that concern exploration, infrastructure development, mine development, and reclamation.

To remove any impediment in the Indian Act and the Indian Act Regulations, in provincial and territorial legislation, in federal-provincial agreements, or under development policies, to full participation of Aboriginal peoples in economic opportunities in mining and related businesses."[142]

Mining's Aboriginal Engagement Progress: BC & YUKON CHAMBER OF MINES – Ten Principles of Sustainable Relationships between the Mineral Sector and First Nations

Dan Jepsen, Executive Director and Chief Operating Officer of the BC & Yukon Chamber of Mines (BCYCM), presented the first Working Draft of this Guidebook at the Aboriginal Engagement & Sustainability Conference in Vancouver on February 8, 2005.[143] During his presentation Mr. Jepsen outlined the "10 Principles of Sustainable Relationships between the Mineral Sector and First Nations" (the 10 Principles):

i) "recognize the asserted traditional territories and areas of cultural or heritage interest of Aboriginal Communities;
ii) recognize that Aboriginal Communities have overlapping or shared traditional territories;
iii) support the conclusion of fair, affordable and reasonable treaties (support the respect, recognition, and reconciliation process between Canada, British Columbia, Aboriginal People and First Nations);
iv) respect the diversity of interests and cultures among Aboriginal Peoples;

142 *THE WHITEHORSE MINING INITIATIVE, LEADERSHIP COUNCIL ACCORD*, Final Report, October, 1994, 27
See: www.natural-resources.org/minerals/csr/docs/guidelines/UNDESA-UNEP%20Guidelines/App%206%20-%20Whitehorse%20Mining%20Initiative.pdf
For a good review of WMI, its history, implementation progress, and future prospects four years after the signing of the Leadership Council Accord, see:
A PRESENTATION TO THE HOUSE OF COMMONS STANDING COMMITTEE ON NATURAL RESOURCES AND GOVERNMENT OPERATIONS, by Gisele Jacob, Vice-President, Public Affairs, The Mining Association of Canada, November 5, 1998
www.mining.ca/english/publications/wmi.htm

143 *BUILDING SUSTAINABLE RELATIONSHIPS, Aboriginal Engagement & Sustainability Conference*, February 8-9, 2005, hosted by Canadian Business for Social Responsibility (CBSR).

v) respect the internal affairs of Aboriginal Communities;
vi) have a common commitment to sustainability and respect for the land and its resources;
vii) recognize that Aboriginal Communities have varying interests and objectives in relationships and cooperative ventures;
viii) acknowledge that there is a shortage of capital to involve Aboriginal bands in cooperative ventures;
ix) encourage the enhancement of Aboriginal capacity to develop training, employment and business opportunities in the resource sector; and
x) support Aboriginal aspirations in securing economic development."[144]

Noting that the 10 Principles were originally developed in consultation with 32 coastal First Nations and Western Forest Products Ltd., and were subsequently edited and adopted by BCYCM to guide its positions on all activities and issues related to Aboriginal communities, Mr. Jepsen emphasized that the strength of their impact potential depends on commitment:

> "Adopted and committed to at a very high level, the 10 Principles can lay the foundation for an organization's positive relations with First Nations. These principles are more than words - they signify commitment to values, and must have clear buy-in from all levels of the organization if you expect results."

Part II of this Guidebook is intended to reflect the spirit of both the Whitehorse Initiative and the 10 Principles.

The following Case Studies, Pointers, Practical Advice and Best Practices are provided as examples and suggestions, not as templates or recipes. Each situation and each First Nation is unique and should be approached as such. It is our hope and goal that the experiences and principles set out in the following pages will inform and equip the reader to approach Aboriginal community engagement with respect and patience, seeking opportunities to build mutual opportunity and shared rewards. The following words have been used previously by BCYCM to introduce the 10 Principles, and are applicable here:

> "Successful partnerships require commitment, hard work, understanding, trust and mutual respect. They also involve recognizing each other's values and aspirations, and identifying and communicating common goals. The Chamber shares First Nations' goals of respecting land and resources, and conducting activities in economically, socially, and environmentally responsible ways, to ensure long-term sustainability."[145]

[144] See: www.chamberofmines.bc.ca/firstnations.htm
[145] BCYCM, *Statement on First Nations Treaties in British Columbia*, January 2004, 3
See: www.chamberofmines.bc.ca/SiteCM/U/D/C4152CA7DCA61272.pdf

Three Case Studies

Progressive, Profitable Partnerships

Introduction

Set out below are three examples of successful partnerships involving members of the mining and Aboriginal communities. These Case Studies are not Utopian – serious obstacles and challenges have been encountered, and others will be faced in the future. Neither are these demonstrations of turnkey solutions: each Case Study presents a very different scenario, and features unique partnership conditions. Yet, common qualities can be found in the ways the parties approach one another and the process of engagement, including: starting from positions of genuine and mutual respect, developing human relationships, emphasizing communication, sharing opportunities, and seeking mutual solutions.

CASE STUDY 1

Vancouver Island's Eagle Rock Quarry Project:

A Modern Industry-First Nations Mining Partnership Between Polaris Minerals Corporation, the Hupacasath First Nation, and the Ucluelet First Nation

Introduction

Eagle Rock Materials Ltd. (Eagle Rock) is a private B.C. company, formed in 2002 to hold the interests of a partnership between Polaris Minerals Corporation, the Hupacasath First Nation and the Ucluelet First Nation. Eagle Rock's goal is to produce high quality construction aggregates (crushed stone) for marine shipment from Vancouver Island to major construction industry markets along the Pacific coast of North America – primarily San Francisco Bay and the Los Angeles Basin.

The Eagle Rock Quarry has the potential to become one of the largest volume and lowest cost producers of high quality construction aggregate materials on the west coast of North America. The Quarry site holds a high-grade granite deposit of about 710 million tonnes, has deep water port access, and is suitable for long term development with minimal environmental and social impacts. Eagle Rock's long term development plan calls for annual marine shipment of up to 6 million tonnes of premium aggregate material, with the project's lifespan estimated at over 100 years.

In the fall of 2003, Eagle Rock obtained its provincial Environmental Assessment Certificate and Mining Permit, following an extensive public consultation and hearing process that raised no public opposition. Eagle Rock is now moving to secure marketing partnerships, strategic locations and regulatory approvals for California dockside terminals, and to achieve a cost-effective bulk shipping solution to its transportation needs.

The Eagle Rock Partnership Agreement

The Hupacasath, Ucluelet and Tseshalt First Nations are all members of the ancient Nuu-chah-nulth Nation. Their traditional territories include the Eagle Rock Quarry site on the shore of Vancouver Island's Alberni Inlet. In July 2002, Polaris executed a business agreement with the Hupacasath and Ucluelet First Nations, under which Polaris holds a 70 per cent interest in Eagle Rock, the Hupacasath and Ucluelet First Nations each hold a 10 per cent interest, and a further 10 per cent interest is held in trust for the Tseshaht First Nation.[146] The trust interest will be divided equally between the Hupacasath and Ucluelet Nations upon a triggering event if the Tseshaht has not yet joined the partnership. Eagle Rock's First Nation partners are represented on the Board of Directors, and the business agreement clearly spells out the

[146] While the Tseshalt First Nation have been invited to join the Eagle Rock partnership, to date they have declined to do so.

corporation's decision-making processes – particularly regarding environmental issues.

The agreement contains several features that separate this partnership from more traditional industry-Aboriginal ventures. Polaris has agreed to finance Eagle Rock until a construction decision is made. On the twenty fifth anniversary of the project development financing, each First Nation partner will have the one-time right to increase its ownership in Eagle Rock by 50 per cent – potentially increasing First Nations equity in the project to 45 per cent. In the event that treaties are settled over the quarry area, the First Nation partners have agreed that, for a term of at least 25 years from the date of such treaties, they will not impose a tenure or tax regime on Eagle Rock with terms less favourable than what would apply without such treaties. Finally, an Impacts and Benefits Agreement has been negotiated, providing Eagle Rock's First Nation partners with preferential opportunities for business development, employment and training, and direct community funding.[147]

Keys to Putting the Eagle Rock Partnership Together

Marco Romero, Polaris's President and CEO, believes that his company's adherence to the following Aboriginal engagement principles was crucial to the successful conclusion of the Eagle Rock partnership negotiations:

- respect for the asserted traditional rights and claims of its First Nation partners;
- acknowledgement that those rights include unextinguished Aboriginal rights, and potentially, title;
- early and sustained engagement by senior management in communication and relationship-building;
- impartiality in dealing with its current and potential First Nation partners, and refraining from taking positions on issues between them – particularly issues involving their overlapping claims to traditional territories; and
- recognition that First Nation communities generally lack the capacity (expertise and capital) required to evaluate complex resource development opportunities.[148]

In a recent interview, Mr. Romero said:

> "Polaris approached its potential Eagle Rock partners very early in the process of research and preparation. We asked for permission to conduct field research on their traditional territories and to meet with them to explain our project concept. We encouraged them to engage experts and consultants of their choice to evaluate the information we presented to them, at our expense. We answered all their questions and provided all the information they requested."[149]

147 Polaris Minerals Corporation Annual Report 2002, 3

148 Interview with Marco Romero, November 19, 2004: Polaris encouraged its First Nation partners to agree to share equally in all aspects of project participation, pending resolution of their overlapping claims.

149 Supra.

Judith Sayers, Chief Councillor of the Hupacasath First Nation, confirmed the importance of Polaris's early and open introduction of the project proposal:

> "Polaris showed us respect and patience. They included us in research and planning from an early stage of project concept development. They gave us unlimited information, funded our due diligence, and answered our endless follow-up questions. We played an important role in developing the core concept and the standards applied to the project. Our fundamental priorities of protecting the environment and creating capacity for our community were laid on the table, and Polaris worked with us to build those priorities into the parameters of the Eagle Rock project."[150]

Chief Councillor Sayers noted the fairness Polaris has shown in its negotiations and on-going dealings with its First Nation partners:

> "From the beginning and throughout, Polaris has dealt with all three First Nations fairly, equally and openly. Respect and trust for Polaris and within the partnership has been strengthened by that approach. We still have to resolve the territorial overlap between the three First Nations, but there is far less impact on the project when Polaris so consistently steers clear of the issue."[151]

Throughout our interview for this Case Study, Polaris's CEO emphasized the scale and duration of commitment to communication and relationship-building required of senior management:

> "Enormous amounts of time and energy have been invested by our management team in an attempt to achieve meaningful consultation. It's all been about good faith engagement over the long haul. We have spent countless valuable and enjoyable hours getting to know our partners, learning their needs and values, fishing, hiking and eating together, and meeting their families with our families. Certainly we've had conflicts and tough negotiations – it took a year to get our partnership agreement done - but we've built a base of trust and mutual respect. More importantly, we've developed friendships."[152]

In discussing strengths of the Eagle Rock partnership, the Hupacasath leader keyed on three aspects of the importance of the early inclusion and active involvement of the First Nation partners:

> "First, it was a huge advantage in the public consultation element of the permitting process. The project quickly gained credibility and support throughout the community and with regulators, because from early on we were publicly standing shoulder to shoulder with Polaris's leaders. Second, our environmental assessment and other permitting processes were simplified by the prior resolution of two of the three First Nations' territorial issues by agreement with Polaris, and by

[150] Interview with Judith Sayers, Chief Councillor of the Hupacasath First Nation, November 25, 2004.
[151] Supra.
[152] November 19, 2004 Interview with Marco Romero, Ibid.

our collective application for the various permits and certificates. Polaris benefited from relationships the First Nations already had in the community with local and provincial governments, and with other key businesses (such as Weyerhaeuser). Third, the active involvement of Aboriginal partners made a positive impression on the senior management of our potential California customers when they came to visit. They were worried about the impact of unresolved First Nations title claims on the certainty of supply of aggregates. If we were involved in the project, our Californian customers knew that the issues of Aboriginal title, consultation and accommodation were far less likely to pose a problem. They valued the contributions to political and public good will, and the business certainty achieved by our committed participation in the project."[153]

Ms. Sayers identified the building of community capacity as the key contribution of the Eagle Rock Quarry project to the Hupacasath community:

> "Once we had confirmed that the quarry concept could be environmentally benign, and we had negotiated a business agreement committing the project to high environmental standards, we quite quickly focused on the potential for strengthening our most important assets – our young people. Eagle Rock's future employment opportunities provide tangible incentive and purpose for our youth to pursue advanced education and skill development. When I talk to our youth, I tell them, "Someday you can be one of Eagle Rock's janitors, its CEO, or anything in between. It's up to you." At least as important as the jobs, our direct involvement in ownership and control of the project builds community pride, creating hope and inspiration about entrepreneurial possibilities for our people."[154]

As an eight year-old boy growing up on the Ucluelet reserve, Tyson Touchie recognized that his people needed to find new resources to survive. At an Eagle Rock Open House held as part of the public consultation element of the permitting process, he explained:

> "Even then, I knew we were in dire straits. We had no oil, we had no gas – we had nothing to sell. I remember saying, 'I hope there's a way of selling those rocks, because that's about all we've got here.' Then, when I came back from school, my Dad said, "Guess what? We're going to sell rocks to the Americans."[155]

Now the Eagle Rock coordinator for the Ucluelet First Nation, Mr. Touchie notes that the project mandate calls for 50 per cent of Eagle Rock's labour force to be supplied by its First Nation partners:

> "For us, sitting on the sidelines is not enough. Eagle Rock is willing to recruit and train young workers.

153 November 25, 2004 Interview with Judith Sayers, Ibid.

154 Supra.

155 Shayne Morrow, ... *Childhood thoughts of selling rock come true for Ucluelet leader*, The Alberni Valley Times, March 6, 2003, 4

If you want to become part of a winning team, this is it."[156]

He shares Judith Sayers's view of the importance of First Nations' involvement in ownership:

"The biggest benefit for First Nations is being acknowledged by industry. This will create a model for future industry to come into our area. And as far as local ownership is concerned, it never occurred to us before – we're as local as you get."[157]

It's not surprising that Polaris's CEO and the Hupacasath's Chief Councillor hold similar views on the business implications of the Supreme Court of Canada's recent decision in the *Haida Nation* case. Neither leader sees the ruling as exempting industry from the duties of consultation and accommodation regarding potential infringement on claims of Aboriginal rights and title. According to Marco Romero:

"It's just good business to keep your neighbours and those who have interests in the land informed, and to give them opportunities to participate in and contribute to the decision-making process. Beyond the question of strict legal duty, we believe that the resource sector has a moral duty to consult meaningfully – not only with First Nations affected by our proposed developments, but also with local non-Aboriginal communities."[158]

And, as Judith Sayers says:

"The smart companies will continue to see consultation and accommodation as building blocks for establishing positive, enduring relationships with Aboriginal communities. Those companies don't waste time and energy worrying about unextinguished rights and title – they've moved way past that, to focusing on the potential for success that lies in building mutual value and certainty."[159]

Eagle Rock offers an exciting glimpse of the future that's possible for Aboriginal participation in mineral exploration and mining development in British Columbia, and throughout Canada.

[156] Supra.
[157] Supra.
[158] November 19, 2004 Interview, Ibid.
[159] November 25, 2004 Interview, Ibid.

CASE STUDY 2

Building Trust and Value Over the Long Haul:

North American Metals Corp. (NAMC) and The Tahltan First Nation - Golden Bear Mine, Near Telegraph Creek, B.C.

The Tahltan's Chief Jerry Asp

Chief Jerry Asp of the Tahltan First Nation understands mining. He cut his mining teeth on the diamond drills back in 1965, working both surface and underground. After two seasons of prospecting the Yukon's St. Elias Mountains, Chief Asp went underground at the Tantalus Butte Coal Mine, near Carmacks, Yukon, where he served as Captain of the Underground Rescue Team and as President of the only all-Native United Steelworkers Local in North America. Moving on, Chief Asp managed Dease Valley Resources Ltd., soil sampling, geo-chem sampling, staking claims, and line-cutting in the Yukon and northern B.C. In 1985, he started the Tahtlan Nation Development Corporation (TNDC) to take advantage of the mining industry's construction opportunities. In 1989, while he was President, TNDC and the Tahltan Tribal Council jointly won the BC Environmental Award. A year earlier, Chief Jerry Asp negotiated B.C.'s first Impacts and Benefits Agreement (IBA) in British Columbia, between the Tahltan Nation and North American Metals Corp., owners of the Golden Bear Mine.[160]

Golden Bear's Innovative Technology and Golden Production

Located about 150 kilometres west of Dease Lake, British Columbia, Golden Bear was one of Canada's first and most successful heap leach gold mines. Its low-grade reserves made the Golden Bear Mine better suited to modern heap-leaching methods than to traditional mining and milling. Fifty tonne truck-loads of ore were dumped onto two huge heap-leach pads, lined with several layers of impermeable black plastic.[161] A cyanide solution was dripped through the ore pile and as it saturated the ore, the gold dissolved into the solution. The 'pregnant' solution was then pumped into a container filled with ground coconut shells. After the gold stuck to the carbon in the shells, the cyanide solution was dripped through the ore pile on the leach pad a second time. The coconut shells were taken to the mill, where under very high temperatures, the gold was separated from the carbon and poured into bars.[162]

Golden Bear was projected to produce 215,000 ounces of gold over six years, at a production cost of US$233 per ounce. In fact, Golden Bear yielded over 265,000 ounces between 1996 and 2001 at a cash cost of about US$170 per ounce, reaching peak production of

160 Chief Jerry Asp's Biography for the Manitoba Mining & Minerals Convention, 2004
See: www.gov.mb.ca/itm/mrd/minerals/convention/speakbio.html

161 The heap leach pads were built literally on a mountain top. During the mine-opening celebration, project manager Randy Smallwood told a mine construction story of a curious ewe sheep that wandered onto the slippery surface of one of the pads. "She couldn't get out", Smallwood said. He went on to explain that a set of stairs had to be built for the use of that ewe and future four-legged visitors. See: www.yukonweb.com/community/yukon-news/1997/sept19.htmld/ , page 16 of 27.

162 Supra.

94,000 ounces in 2000. The mine created about $43 million in cash flow during a period of relatively low gold prices.

The 1998 Golden Bear Agreement

Before beginning negotiations with North American Metals, the Tahltan Tribal Council created the Tahltan Resource Development Policy to establish the ground rules for resource development. As Chief Jerry Asp explained in a speech at the 2003 First Nations Exploration and Development Symposium:

> "The 1998 Golden Bear Agreement ... highlights governance structure and how a mining project may fit into the long-term plans of a community. We weren't looking for a few jobs, seasonal employment or a training program or two ... We needed a plan that would benefit our entire nation, not just a few individuals. In other words, we were looking for a nation-building model for economic development, not a job and training model."[163]

The Golden Bear Agreement guaranteed a number of key benefits for the Tahltan First Nation, including: "work on building the 160 kilometre road into the mine site, a three-year upgrading and maintenance contract, a minimum of 20 per cent of the work force at the mine site and the right to negotiate for future contracts. The Tahltans negotiated building the settling ponds, a five-year open-pit mining contract and general camp maintenance."[164]

It wasn't all smooth sailing. In April of 1997, the Tahltan blockaded the mine's access road, drawing attention to their concerns about being excluded from contracting opportunities and to their request that further environmental assessment be done on the mine's potential impact on their traditional territories. As Yvonne Tashoots (Tahltan Chief at the time) said in an interview with the Yukon News, "It was confrontational at the start." She noted that for several months all communications were filtered through a mediator, and concluded, "But as a result of our communication, we have all learned to respect each other a little more."[165]

By September of that year, Tahltan band members made up 35 per cent of Golden Bear's employees, and the Tahltan had the mine camp's catering contract, as well as the maintenance contract for the road they had blockaded. "Things have improved," said Ian McDonald, (then-CEO and Chairman of Wheaton River Minerals Ltd. – owners of North American Metals). "I think they thought we were ignoring them and, perhaps, we were a little guilty of that."[166] It should be noted that in April, 2003, the Tahltan Mining Symposium was convened to:

163 Jerry Asp, ***Mining ... A Catalyst for Aboriginal Community Development***, MINING REVIEW, An Official Publication of the B.C. & Yukon Chamber of Mines, (Burnaby: Canada Wide Magazines and Communications Ltd., Fall 2003) 8

164 Supra.

165 www.yukonweb.com/community/yukon-news/1997/sept19.htmld/, 15 of 27

166 Supra.

"(1) review the relationship between the Tahltan people, their land and the mining industry; and (2) build a strategy to guide that relationship in the future. Seeking a win-win outcome, and guided by the Seven Questions to Sustainability (7QS) Assessment Framework, the participants considered past, present and potential future conditions as a foundation for ensuring positive outcomes for the Tahltan people and their territory in the years to come."[167]

CASE STUDY 3

Alaska's Red Dog Mine:

Co-operative Development by Teck Cominco, the Northwest Alaska Native Association (NANA) and the State of Alaska

Located 140 kilometres north of the Arctic Circle, the Red Dog Mine opened in 1989 and is the world's largest producer of zinc concentrate. The 94.4 million metric tonne ore body is a rich and highly concentrated deposit of 17 per cent zinc, five per cent lead, and 2.4 oz/ton of silver, in an area of just 1.6 km by 0.8 km – lying within the settlement lands of the Northwest Alaska Native Association (NANA).[168]

In 1982, NANA entered an agreement with Cominco Ltd. (Cominco – now Teck Cominco Limited), giving Cominco rights to build and operate Red Dog and to market its metal production, in exchange for escalating royalties to NANA and a commitment that within 12 years of mine opening, 100 per cent of hourly wage jobs at the mine would be filled by NANA members. The royalty package provided NANA with an initial payment of $1.5 million, a construction period royalty of $1 million per year, and a production royalty of 4.5 per cent of net smelter returns. The production royalty will increase to 25 per cent once Cominco recovers its capital investment, and will

167 For a report on the work of the Tahltan Mining Symposium, see: International Institute for Sustainable Development, ***Out of Respect - The Tahltan, Mining and the Seven Questions to Sustainability***, See www. iisd.org/natres/mining/tahltan.asp (IISD Publications Centre, 2004) The BC & Yukon Chamber of Mines was pleased to support the preparation of ***Out of Respect***.

168 Ron Mclean and Willie Hensley, ***Mining and Indigenous Peoples: The Red Dog Mine Story***, (Ottawa: The International Council on Metals and the Environment, 1994) 1-4

increase by five per cent every five years until NANA's share reaches 50 per cent, where it will remain for the life of the project. [169]

Despite Red Dog's remote location and high development and out-shipment costs, the development's lifespan is projected at 45 to 50 years – a credit to the size and quality of the deposit, and to a unique provision in the 1982 development agreement. Cominco agreed to cap annual production at 2.33 million metric tonnes of ore, regardless of the prevailing price of zinc - stabilizing and extending Red Dog's production and employment.

The State of Alaska's Industrial Development Authority invested $160 million in building and financing the mine's transportation system (83 km. haul road and seaport). The state's investment is being repaid with interest out of Cominco's user and export fees, over the life of the mine.[170]

Cominco's commitment to have NANA members filling all Red Dog hourly wage jobs within 12 years was (and continues to be) a major challenge, given that no one in the community had mining experience in 1982. First, Cominco had to design the Red Dog concentrator to be sophisticated enough to achieve cost-effective recovery of saleable zinc, lead, and silver concentrates from the ore, yet simple enough to be operated by inexperienced milling crews. Second, an on-going commitment to employee recruitment, scholarship programs to support education, and technical training programs has been required. A 26-week pre-apprenticeship program was originally offered in the city of Seward, but high dropout rates led Cominco to relocate the program to the village of Kotzebue (the community closest to the mine site).

Employment dropout rates in the range of 20 per cent in the early years of Red Dog's operation were attributed largely to conflicts between shift scheduling and subsistence activities. Seasons are short and inflexible for berry-picking and hunting, so shift rotations were adapted to accommodate employees' traditional pursuits. By 2000, the employment dropout rate had improved to 11 per cent. By July 2002, NANA members held 56 per cent of all Red Dog-created jobs. Flexible shift rotations and continued commitment to training programs are expected eventually to fulfill Cominco's commitment of 100 per cent NANA sourcing of hourly wage positions. [171]

169 Supra, 7
170 Supra, 6
171 Julie Domvile, ***RED DOG MINE: A Role Model in Co-operative Development***, in "MINING REVIEW", Ibid, 14-15

Environmental protection is a major priority for Red Dog, Cominco, and NANA. In addition to a comprehensive Cominco Environmental Policy, the Red Dog Mine Operating Policy contains a strong environmental protection clause:

"PROTECTION OF THE ENVIRONMENT

- To comply with all existing local, state and federal laws and regulations. To provide additional environmental protection measures, where warranted, that are technically feasible and economically viable.

- To encourage, support and conduct necessary research to establish high standards of performance and to improve methods for environmental control. To keep employees, agency personnel, and the general public fully informed concerning the environmental aspects of company operations. **In all emergency situations protect in order of priority: personnel, environment, property and production."**[172]

Even more telling is the purpose, scope, and 'on-the-ground' power of Red Dog's Subsistence Committee. Created pursuant to the terms of the 1982 NANA/Cominco Operating Agreement, the Subsistence Committee is made up of elders from the villages of Kivalina and Noatak, and meets quarterly or as needed:

"The prime purpose of the Subsistence Committee is to ensure that all exploration, development and mining activity at the mine site is consistent with sound stewardship principles and will not harm or threaten the subsistence needs and the physical, cultural, social and economic needs of the indigenous people of the NANA region. The committee reviews many reports from extensive environmental monitoring required by Cominco and by the government permits. The quality of water, air and earth is continually tested. Any possible effects on the two nearby villages are openly discussed. Activity on the 83 kilometre concentrate haul road from the mine site to the Port site is monitored, and potential effects on subsistence hunting are reviewed by the committee.

During the caribou migration season, the Subsistence Committee can shut down traffic on the road, especially if the caribou are crossing in large numbers. In addition to stringent regulatory and policy requirements, the local people who live in the area have authority to close down the road if they legitimately feel the impact is detrimental to their subsistence."[173]

The Red Dog Mine[174] is a strong demonstration of what can be achieved by combining state of the art

172 Mclean and Hensley, ***Mining and Indigenous Peoples: The Red Dog Mine Story***, Ibid, Appendix B, 24

173 Supra, 9

174 Supra, 4. The Red Dog deposit was discovered in 1953 by Bob Baker, a local bush pilot and prospector, who noticed the rusty alteration of Red Dog Creek as he flew over the area. He reported the find to Mr. Irving Tailleur of the U.S. Geological Survey, who was mapping the geology of the DeLong Mountain quadrangle which includes Red Dog. When Mr. Tailleur visited the area, he "immediately noticed abundant barite, black chert, siliceous sinter, and iron oxide staining. His few rock samples showed significant zinc and lead mineralization.

Mr. Tailleur's findings and the apparent similarities of this occurrence to other large zinc/lead deposits around the world were documented in a USGS open file report published in 1970. The name "Red Dog Creek" was coined by Mr. Tailleur, after Bob Baker's prospecting company, the Red Dog Mining Company, named after his pet dog which frequently flew with him."

technology with an imaginative, open approach to negotiation and partnership structuring. Modern mining can be compatible with a fragile ecosystem and the subsistence needs of Aboriginal partners.

POINTERS, PRACTICAL ADVICE & CORPORATE BEST PRACTICES:

Elements of Sustainable Relationship Between the Mineral Sector and First Nations

INTRODUCTION

Mineral Exploration: Mining's 'First Impression'

Mineral exploration activities may have powerful social and cultural dimensions, whether or not the crews on the ground are aware of the broad impact of their presence and actions. Exploration personnel usually make the industry's first contact with a local community. In a very real sense, exploration companies and their crews are ambassadors for the entire mining sector, and the first impressions they make will colour community perceptions and responses toward mining in the future. **First impressions count and are long lasting – you don't get a second chance to make a first impression!**

Communities often interpret the arrival of heavy equipment (i.e. road-building machinery and drilling rigs) as the commencement of mining operations. The highly speculative and preliminary nature of mineral exploration is not widely understood outside the industry. BCYCM experience suggests that the odds against a mineral exploration project being developed into a producing mine are in the range of 1,500-1.[175]

The exploration team must be aware of the risks created by unrealistic expectations, and must work hard to provide the community with accurate, transparent, and timely information. Sometimes exploration teams will suggest, or even promise that benefits will flow to individuals or entire communities when mines go into production. Most often, these are promises of future employment or business opportunity. But, those exploration teams or companies, particularly in the case of junior companies, are seldom around if and when the mine finally opens. Unfulfilled promises lead to frustration, resentment, and conflict. Exploration teams should not make promises that they can not guarantee will be kept.

The exploration phase of the mine cycle can last from two to twenty years, or more, during which the Aboriginal and local communities may be exposed to many different companies and individuals. The attitudes of those communities towards mining and mining companies will be strongly influenced by the social and environmental performance of these exploration companies. The nature of the local perceptions and opinions created during exploration will strongly influence Aboriginal communities' willingness to welcome a mine into their traditional territories. For an exploration or mining company to be

175 Michael Farnsworth, Chair of BC & Yukon Chamber of Mines' First Nations and Community Relations Committee, in an Interview, December 21, 2004

welcomed by an Aboriginal community, deeper qualities than protocol and courtesy are required. Aboriginal Peoples have learned to be wary of non-Aboriginal visitors. Their trust must be earned over time, by demonstrating genuine respect and care for the land, its plant and animal life, and the people whose ancestors have lived on that land for thousands of years.

Context for Aboriginal Community Engagement

The business case

Early and effective Aboriginal community engagement in mineral exploration and mining projects makes good business sense for communities and sector companies. Benefits include:

- Quicker Permitting Processes
  - Facilitated Government review and permitting process, with the proponent completing some of the practical on-the-ground aspects of consultation; strengthening Government decision to permit
  - Increased community participation in the project, ultimately leading to greater support
- Greater Community Support
  - Increased community awareness of project activities and the minimal footprint of mineral projects compared to some other industrial activities
  - Greater community comfort with project activities
  - Improved access to Aboriginal community's labour pool
  - Enhanced quality of community life through increased employment

The legal case

We have already reviewed the line of Supreme Court of Canada decisions that declare and define the **Crown's legal obligation to consult** with Aboriginal groups to ensure that Government actions and decisions do not unjustifiably infringe upon Aboriginal and treaty rights, whether claimed, proven, or settled by treaty. While the Supreme Court recently confirmed in *Haida Nation*[176] that the duties of consultation and accommodation sit exclusively with the Crown and may not be delegated to third parties, project proponents should understand that:

- Government consultation is **not a duplication of company engagement effort**
- Company-community engagements complement Government consultation
- To make informed permitting decisions, Government staff need to be advised of company-community engagements, including consultations and any resulting agreements and accommodations, such as training programs and project-related jobs
- If Government consultation and accommodation measures are determined to be inadequate, permits and approvals may be set aside and pending projects may face delay or even cancellation

[176] *Haida Nation*, Ibid, see pages 43-44 for discussion

- **Relying on Government to perform consultation and accommodation on industry's behalf is not prudent business practice**[177]

POINTERS

- Watch for legal requirements for advanced projects (i.e. permit-related activities such as road building and drilling) that have potential to infringe Aboriginal Rights or Title (whether claimed or confirmed)
- Commit your organization and your people
  - To common courtesy
  - To honesty and integrity in all your dealings
  - To open and consistent communication
  - To maintaining good business practices
  - To remembering that credibility and trust
    - are built over the long haul, and
    - are lost in a day
- Build good working relationships with First Nations
  - To boost public trust and support
  - To spark market interest
  - To enhance regulatory and environmental movement confidence
  - To realize business certainty
- Inform locals of company plans and ensure their understanding of them
- Help manage expectations/rumours about proposed activities
- Communicate regularly regarding
  - activities of the company
  - the status of the project
  - current and future opportunities for net benefits from the project
    - To earn community trust and good will
    - To enhance local understanding of and support for the project and related activities
- Recognize that Aboriginal people and local communities have a substantial influence on project approval and successful development (not only in British Columbia and the rest of Canada, but increasingly around the world)
- Recognize that "consultation" is more than an element of public relations or due diligence, and entails more than developing good relations with Band leadership
  - "Consultation" entails communication with the entire Aboriginal community(ies) likely to be affected by a development, and requires ensuring that a broad and inclusive representation of the community(ies) is briefed on the proposed project's expected and potential community impacts
- Learn of First Nation concerns, issues and thoughts about the project and consider how to deal with them
  - Find out what band leadership and the community see as key heritage issues
    - Ask, "What do you wish for your children?" and pursue discussion around that theme
- Learn of availability of local labour and services
- Learn of need for and availability of training programs
- Seek opportunities to develop joint

177 "Businesses in Canada will ... be well-advised to ensure that whatever consultation or accommodation is required by the Crown is conducted at the now legally required standard because they will ultimately pay the price of inadequate consultation in questionable permits, licenses or other approvals." Tom Isaac and Tony Knox, *The Crown's Duty to Consult and Accommodate Aboriginal People: The Supreme Court of Canada's Decisions in Haida Nation and Taku River Tlingit*, Ibid, 7

initiatives and to make joint presentations with local First Nations

Before any Contact

- Collect available information regarding consultation and local First Nations (The BC Treaty Commission is a good resource: www.bctreaty.net)
  - Including asserted territory maps, First Nation websites, affiliations, etc.
- Learn locations and names of all interested First Nation communities
- Learn locations of all nearby Indian reserves
- Make sure that those reserves are strictly off limits to entry or transit without permission
- Make sure that all crews, contractors and suppliers are notified of those access restrictions
- Learn any Tribal Council or other political affiliations of interested First Nations
- Learn names of all contacts (First Nation Band Chiefs, Hereditary Chiefs, Administrative Officers, etc.)
- Learn applicable decision making/governance structures and the roles of Hereditary and Elected Chiefs and Elders
- Learn community priorities
- Learn locations of any local areas of cultural significance or other sensitivity
  - Be aware that some Aboriginal communities may be reluctant to disclose the location and other information regarding such areas
- Learn accepted and asserted boundaries of Traditional Territories of local First Nations
- Learn any local history and in particular any negative aspects of local First Nations' past mining/exploration experiences
- Learn who has influence in the community
- Identify and learn the important and sensitive local issues
- Understand the Aboriginal consultation procedures used by the applicable orders of Government in the review and permitting process for mineral projects
- Be aware that for projects at the permitting stage, provincial permitting staff will be consulting with Aboriginal communities in order to fulfill Crown obligations
- Understand that such governmental consultation is **not a substitute for or duplication of company engagement effort** (please see pages 80-81 for discussion)
- Understand that in order to make informed permitting decisions, and to discharge the Crown's duties of consultation and accommodation[178], Government staff need to be advised of company-community engagements, including
  - Consultations
    - Any resulting agreements and accommodations, such as
      - Training programs and project related jobs
  - Conflicts and disputes
  - Incidents involving trespass or damage to Aboriginal territories
    - whether claimed or confirmed

[178] See pages 40-46 for discussion of the *consultation and accommodation* issues arising from the recent Supreme Court of Canada decision in the *Haida Nation* case, Ibid.

- Understand that industry failure to provide such permit-sensitive information, and government failure to discharge the Crown's duties of consultation and accommodation may expose your operation or development to future legal consequences that may jeopardize your permits and land access. (Please see pages 80-81 for discussion).
- Become familiar with any existing training and employment initiatives that might be available to your company and interested Aboriginal communities

Sources of Information

- Local books, museums and news media
- Community Cultural Centres
- First Nations web-sites
- Government representatives
- Band economic development officers
- Consultants and lawyers
- Local suppliers and contractors
- Other companies working in the area
- For advanced and ongoing programs/projects, consider
  - Retaining an Aboriginal Community Relations Officer
    - Accountable to CEO or other senior executive
    - Responsible for developing, maintaining and enforcing corporate policy and procedures for communications and negotiations with Aboriginal groups
  - Retaining an Aboriginal Employment Liaison person
  - Hiring and training Aboriginal Environmental Monitoring personnel

Making Contact

- Make all initial contacts in person, along with assistants, advisors, etc.
- First contact should be made by the senior company personnel
  - Not consultants
  - Not lawyers
  - Later conversations can move from Chief-to-Chief to Advisor-to-Advisor
- Persistence and patience are essential
  - Your time line is not <u>their</u> timeline
  - Rushing can lead to challenges, conflicts and resistance
- Be aware that meetings can be disrupted or delayed – by
  - Traditional harvesting activities (berry picking, hunting or fishing)
  - Unexpected events such as a death in the community
- Check for harvesting times and be sure to re-confirm meeting status up until the last minute
- Communicate and act with respect, courtesy, transparency and openness
- Be consistent in approach and living up to commitments
- Be prepared to meet people from several Aboriginal communities, although
  - Be cautious when dealing with different groups in the same meeting
- Initial contact should be at the leadership and decision-making level, with
  - Chief
  - Chief Councillor, Council Member, Band Administrator, and/or Economic Development Officer
- Follow-up with Band or First Nation-designated contact

- Do community 'walk-about' visits
  - Chat or have coffee with Elders and other band members
- Strive to engage with the full community
- Avoid traditional "cold" business form letters and correspondence
- Ensure that sincere efforts are made to consult with the entire Aboriginal community, not just its leaders and officials
- Also contact the local non-Aboriginal public in the vicinity of the project, particularly in the case of advanced projects

What to Say

- Express thanks for being welcomed to their Traditional Territory
- Acknowledge that the First Nation
  - Has resided in this region since 'time immemorial'
  - Has asserted Aboriginal Right, Title and Interest to such territory
- Acknowledge that the Company
  - Hopes to explore on the First Nation's Traditional Territory
  - Hopes to avoid or minimize infringements of Aboriginal Rights, Title and Interest
- Explain the Company's vision for the region and for the proposed project
  - Outline the exploration concept/method
  - Identify the approximate location(s) for proposed exploration
  - Set out the tentative time line(s) for project development.
  - Distinguish between
    - early stage (grassroots) exploration projects, and
    - middle to long term development projects
  - Enable communities to keep a suitable reviewing perspective
- Provide a brief industry overview
  - Including the nature, roles and capacities of the various players, including
    - Prospectors
    - Junior (exploration) companies
    - Senior (mining) companies
  - Minimize the use of technical maps, plans, and reports at this stage, to avoid
    - Overwhelming local people
    - Giving the mistaken impression that plans are already finalized
- Ensure that regional and local scale project location maps with outlines of accepted / asserted traditional territories are used
  - So that the Aboriginal community
    - Can determine if the project is relevant to them
    - Identify any Aboriginal rights in the proposed work areas
- Outline the type of project being proposed
  - Identify minerals sought and type(s) of deposits identified and projected
- Outline projected labour requirements and procurement logistics
  - Describe size, skills and projected duration of labour force
  - Describe shift rotation, worker transportation and accommodations being proposed for the labour force
  - Describe supply contracts being tendered

  - Identify company contact for hiring or contracting
- Outline potential environmental issues and proposed mitigation measures being considered
- Outline sustainability elements of project planning
- Seek feedback and comment from First Nations at all stages, particularly regarding
  - Environmental issues and priorities
  - Sites or concerns involving heritage, cultural or traditional use of which the company may be unaware
- Listen attentively, proactively and positively
  - Watch for clues from body language and group dynamics
- Be prepared[179]
  - To change project plans to address First Nation concerns
  - To provide financial support for First Nation project review and due diligence, particularly regarding advanced projects
  - To accommodate First Nation community needs for further study, discussion and possible program adjustments, thereby
    - Alleviating concerns
    - Reducing opposition
    - Generating support

Information Needed from Local First Nations

- Local work force background
  - Skill and experience
  - Availability
    - Seasonal and subsistence issues
    - Competing and conflicting project issues
- Local business community background
  - Capabilities, capacities and aspirations
  - Availability
    - Seasonal and subsistence issues
    - Competing and conflicting project issues
- Determine level of awareness and exposure to past exploration and mining activities
- Determine whether there were any particularly negative or positive experiences with previous exploration companies and mine operators
  - Obtain details of such experiences
  - Prepare to demonstrate your project's respective contrasts and similarities to those experiences
- Identify local concerns about possible environmental impacts or disruptions of traditional land use activities such as fishing, hunting, berry picking, and cultural events
- Identify local concerns about possible social and community impacts of exploration activity and resource development
- Identify the community's 'opinion leaders' and 'influence wielders'
- Identify the community's important issues
- Assess the community's political dynamics, including

179 See: Case Study 1: Vancouver Island's Eagle Rock Quarry Project: A Modern Industry-First Nations Mining Partnership Between Polaris Minerals Corporation, the Hupacasath First Nation, and the Ucluelet First Nation for examples and discussion. Pages 68-72

- Traditional alliances and enmities
- Personality and issue-based coalitions and conflicts
- Decision-making process and time frame

• Confirm availability of Traditional Ecological Knowledge and other local knowledge

• Determine level of awareness and interest in future Impact and Benefit Agreements (IBAs) if project progresses to more advanced phases

Possible Outcomes

• Establish contact
  - Start getting to know each other
  - Start developing a relationship based on respect, understanding, trust, and honour

• Agree to hold future meetings
  - Develop agenda outlines
  - Establish tentative schedule

• Jointly develop appropriate procurement protocols as applicable[180], addressing
  - Labour force issues, including
    - ❍ Hiring policy and employment policies
      - ❑ Preferred Aboriginal opportunities
    - ❍ First Nation community skill development and training
      - ❑ Bursaries and Scholarships
    - ❍ Seasonal and subsistence shift scheduling issues
  - Local business opportunities
    - ❍ Identify needs, including
      - ❑ Projected product and service needs
      - ❑ Product quality and service performance requirements

• Jointly prepare a Memorandum of Understanding (MOU) to document the consultation that has occurred and what has been decided

• Discuss possible future Impact and Benefit Agreements (IBAs) and provision for transition to more advanced programs (advanced exploration, bulk sampling, feasibility studies, construction, operations, etc.) if they become warranted

Possible Discussion Items for Future Meetings

• First Nation training and hiring issues, including
  - Joint funding applications for government training support
  - Company to assist First Nation(s) with cost of attending Mining Industry Training Courses
  - Company's best efforts to employ First Nation members

• First Nation contracting and subcontracting issues, including
  - Company's best efforts to contract with suppliers owned/controlled by or affiliated with the First Nation
  - First Nation Liaison Person
  - Contractors also to use best efforts
  - Company to submit names of successful contractors
  - Company to advise timing of work of proposed programs and provide lists of jobs and contracts

180 See: Case Study 3, ***Alaska's Red Dog Mine: Co-operative Development by Teck Cominco, the Northwest Alaska Native Association (NANA) and the State of Alaska***, pages 75-78, and Mclean and Hensley, ***Mining and Indigenous Peoples: The Red Dog Mine Story***, Ibid, for excellent review of progressive employment policies and practices in a very demanding environment.

- Potential exploration phase contracting opportunities, including
  - Road work
  - Earth moving
  - Snow clearing
  - Line cutting
  - Geochemical sampling
  - Environmental sampling
  - Camp construction
  - Catering
  - Expediting
  - Fuel supply
  - Reclamation
  - Assisting in environmental assessment, stream-sampling, fish and wildlife assessment, and archeological overview assessment
  - Minimize impacts on fish, wildlife, and water
  - Land interruption fees (general to the First Nation, corporately)
  - Trap line interruption fees (may be specific to individual Band members)
  - Document the discussion and any agreement reached with a Memorandum of Understanding (MOU)
  - For more advanced exploration and pre-feasibility work, consider an
    - Impacts and Benefits Agreement (IBA)

IBA Discussion Issues

- First Nation preferential hiring
- Training and commitment to promote qualified First Nation members
- Company support for scholarships and bursaries
- Employment rotation to accommodate First Nation needs and preferences
- Housing, food service, and recreation
- Safety, health, and hygiene
- Workplace language policy
- Subcontractors' obligations
- Labour relations
- Preferential First Nations contracting policies and practices
- First Nations notification of business and contracting opportunities
- Provision of seed capital and equity/credit to First Nation suppliers, service providers, and contractors
- Land disruption payments
- Production payments
- Support for government resource revenue sharing discussions
- Provision of expert advice
- Information flow and interpretation, including
  - Liaison between First Nation and Company regarding project management and First Nation participation
- Identification, protection, and conservation of cultural, heritage, and archaeological sites and artifacts
- First Nation involvement in
  - Baseline environmental studies
  - Identification of heritage and archaeological sites
- Research and development
- Co-ordination with other area developments
- First Nation concerns regarding environment protection and wildlife disruption, including wildlife/habitat disruption compensation schemes

- First Nation access to facilities provided for the development, including
  - Airstrips, docks, and roads
  - Health and recreation facilities
- First Nation confirmation of adequacy of consultation and accommodation measures
- Arbitration and amendment provisions

Practical Advice

Some Do's and Don'ts

Here are some things to do and not to do when working with Aboriginal People:

Do's

- Research the community and governing parties before going to the community
- Take training on "Working Effectively with Aboriginal People" or attend a similar workshop before you start
- Express your thanks for the invitation into their traditional territory. For example, *"I would like to thank the ___________ First Nation(s) for agreeing to meet with us and inviting us into your traditional territory."*
- Use caution when shaking hands
  - The typical North American elbow grab and double pump may not be needed or appreciated
- Be aware of the importance many Aboriginal cultures attach to body language and non-verbal communication
  - Expect to be observed very closely in everything that you do
- Commit to building a real working relationship
  - Seek to make contact and to meet at a time when you don't need something
  - Have Company management spend some unstructured time in the First Nation community
  - Visit the coffee shop
  - Chat with Elders and other members of the community
  - Ask First Nations how **they** want to be consulted
  - Ask First Nations to define **their** needs and expectations
  - Approach issues with a **joint** problem-solving perspective
- Know the difference between a Band Chief and a Hereditary Chief before you go to a First Nation community
- Be prepared to meet both Band Chiefs and Hereditary Chiefs on the same day and in the same meeting
- Be prepared to say that you are having a problem and that you are there to get some thoughts from them on how to solve it
- Avoid 'Power Suit' formal dress for community meetings
  - Band offices generally have more casual dress policies than those of corporate Canada
- Anticipate questions that First Nation(s) may have of your organization and prepare answers to those questions
- Honour all your agreements, especially your oral agreements
  - Traditionally, these are oral societies
    - Oral agreements are as important in Aboriginal communities as written agreements

- Be flexible and patient
  - Understand that it is not uncommon for the Band office to close on very short notice for various reasons; for example
    - A wedding or a death in the community
  - Call ahead to confirm your meeting time

Don't

- Say that you are there to seek their First Nation **stakeholder** input
- Say that you have a time line and that they have to meet it
- Tell the First Nation what dates are available in your calendar for a meeting
  - Instead, ask which dates would work best for them
- Go to them with a completed draft plan for your project before consultation has started
- Fly in on the Company helicopter at 9:45 for a 10 o'clock meeting, and out again at 11:45 to make your next meeting
- Expect that a consultation approach used successfully with a particular tribe for a particular issue will necessarily succeed in subsequent negotiations with that tribe on other matters (even if the issues are similar)
- Confuse potluck with potlatch
- Confuse reservations with reserves
- Refer to "Indians" or "Natives"
  - Instead, use "Aboriginal People" or "First Nations"
- Say "Some of my best friends are ..."
  - "... Aboriginal People"
  - "... Indians"
  - "... belong to First Nations"
- Ask if they know well known First Nations personalities; for example
  - Chief Dan George
  - Sitting Bull's descendants
- Say that you prefer a municipal style of government
- Say, "We should all be equal."
- Ask, "Are you going to be Canadian when this is all over?"
- Need or expect direct eye contact
- Feel that you must answer or interject whenever there's an interlude in the conversation. Aboriginal cultures often use silent moments in conversation as opportunities to formulate thought
  - Instead, focus on ensuring that the speaker has finished before you contribute to the conversation

Corporate Best Practices: A Checklist

The following checklist has been reproduced courtesy of Pam Sloan and Roger Hill, from their book ***Corporate Aboriginal Relations***[181]. While oriented to larger corporations, this checklist also sends valuable messages to the junior exploration sector, the small and medium-sized business community, and the public at large.

Build Organizational Commitment

Develop and Adopt a Formal Aboriginal Policy

- Appoint a policy development coordinator/project team
- Establish a business case for Aboriginal Relations
- Develop a policy in conjunction with Aboriginal and other stakeholders
- Obtain confirmation that the policy accords with the present state of the law
- Obtain board level approval for policy
- Obtain endorsement/support of union
- Hold formal launch to adopt the policy
- Communicate the policy to internal and external audiences
- Sustain senior management role in promotion of the policy

Allocate Resources to Aboriginal Relations Initiatives

- Appoint an Aboriginal Affairs Coordinator
- Create and staff an Aboriginal Relations Unit (if program size warrants)
- Define mandate and reporting relations to senior management
- Define linkages to and responsibilities of rest of organization

Integrate Aboriginal Relations into Business Planning Process

- Set realistic long-term goals
- Allocate responsibility for goals among relevant business units
- Set annual targets at business unit level
- Establish accountability framework
- Establish monitoring and reporting system
- Integrate accountability into performance review process for managers

Build Knowledge and Understanding about Aboriginal Relations

- Communicate policy/program/results to employees on an ongoing basis
- Communicate policy/program/results to Aboriginal stakeholders on ongoing basis
- Communicate policy/program/results to external stakeholders on ongoing basis
- Provide Aboriginal awareness training for managers
- Provide Aboriginal awareness training for employees

181 Sloan, Pamela & Roger Hill, CORPORATE ABORIGINAL RELATIONS: BEST PRACTICE CASE STUDIES, (Toronto: Hill Sloan Associates Inc., 1995)

EDUCATION AND TRAINING

Encourage Young People to Stay in School

- Establish relationships with schools with large Aboriginal populations
- Provide information to students about education and career options
- Provide role models and mentors for Aboriginal students
- Develop educational experience programs
- Offer short-term work experience programs to students

Provide Education/Training Opportunities and Support

- Develop/support access programs for post-secondary education
- Provide educational awards/scholarships in relevant disciplines
- Develop access programs for employer-based training programs
- Develop/support access programs for skilled trade training programs
- Provide information on skilled trade qualification/certification process
- Develop/support training programs in high skill occupations

Offer Pre-Employment Programs

- Develop/participate in job-readiness training programs
- Link training to concrete job opportunities
- Guarantee employment for successful program participants
- Offer employment skills workshops
- Provide summer employment opportunities
- Offer short-term work assignments

ENHANCING EMPLOYMENT OPPORTUNITIES

Develop Aboriginal Employment Strategy

- Establish advisory committee or special task force
- Ensure participation of Aboriginal and other stakeholders
- Set long-term goals for Aboriginal employment
- Set annual targets for Aboriginal employment
- Identify barriers to Aboriginal employment
- Determine necessary modifications to corporate human resource policies
- Define special measures needed to improve employment opportunities
- Develop action plan
- Establish accountability framework

Target Recruitment Initiatives

- Target outreach activities to organizations with high Aboriginal populations
- Visit Aboriginal communities and training institutions
- Encourage/facilitate visits by Aboriginal people to local offices/workplaces
- Network with Aboriginal education and employment counsellors
- Form partnerships with Aboriginal communities and service organizations

- Keep Aboriginal organizations informed about job opportunities
- Use available inventories of Aboriginal candidates
- Promote development of inventories of Aboriginal job candidates

Facilitate Access to Entry-Level Positions

- Ensure Aboriginal representation in recruitment pool
- Include Aboriginal people in selection process
- Use Aboriginal internship programs
- Negotiate hiring preferences in collective agreement
- Negotiate apprenticeship opportunities in collective agreement
- Create in-house training positions in technical occupations
- Create in-house training positions in skilled trades

Encourage Career Development

- Provide in-house basic education and literacy program
- Adapt corporate training programs to ensure cultural sensitivity
- Ensure Aboriginal access to management/supervisory development opportunities
- Implement succession planning and ensure Aboriginal representation
- Promote and facilitate mentoring

Create a Positive Working Environment

- Implement measures to eliminate harassment, discrimination, and racism
- Promote both the social and business benefits of workforce diversity
- Use Aboriginal employee advisory groups as resource to management
- Facilitate development of Aboriginal employee support networks
- Encourage buddy systems for new Aboriginal employees

ABORIGINAL BUSINESS DEVELOPMENT

Develop Procurement Policies

- Set long-term goals for Aboriginal business participation
- Set annual targets for Aboriginal business participation
- Adopt procurement policies that target Aboriginal suppliers
- Set aside some contracts exclusively for Aboriginal business
- Allocate set asides through negotiated/restricted tendering processes
- Provide long-term supply contracts to promote business formation/expansion
- Require major suppliers to provide opportunities for Aboriginal participation

Remove Procurement Barriers

- Break contracts into smaller packages to provide access to small business
- Ensure that Aboriginal contractors are on bid lists
- Provide early notice to Aboriginal communities/businesses on upcoming contracts
- Clarify procurement processes for Aboriginal businesses

- Use pre-qualification process to promote competitiveness of bids
- Inform potential bidders about legal/safety/regulatory requirements
- Assist potential bidders to meet legal/safety/regulatory requirements
- Waive bid and performance bond requirements, if feasible

Promote Supplier Development

- Develop inventories of local Aboriginal contractors and businesses
- Help Aboriginal businesses to compete effectively for contracts
- Foster development of management skills in Aboriginal businesses
- Encourage joint ventures with non-Aboriginal business to build capacity
- Provide subsidies and financial assistance
- Debrief unsuccessful bidders to help improve future bids
- Create joint opportunities to promote development of large Aboriginal suppliers

Enter into Cooperative Business Ventures

- Enter into cooperative business ventures with Aboriginal organizations
- Structure business ventures to ensure substantive benefits for Aboriginal partners
- Partner with Aboriginal development agencies to develop commercial complexes

COMMUNITY RELATIONS

Establish Communications with Local Communities

- Establish a community liaison committee
- Establish Aboriginal advisory councils
- Consider representation of community leaders on Board of Directors
- Provide ongoing information through Aboriginal affairs group and other staff
- Provide communication materials in variety of formats and local languages

Make Resources Available for Community Development

- Assist local communities to define their community development needs
- Assign staff to work with community on economic/business development strategy
- Open up corporate training courses/workshop to community representatives
- Loan equipment for community projects
- Provide funding for community infrastructure
- Sponsor and promote community events and projects
- Allocate corporate donations to Aboriginal communities
- Promote recognition of Aboriginal achievement in broader community

Develop Collaborative Initiatives

- Establish joint planning and decision-making mechanisms
- Establish joint problem solving/ grievance resolution processes
- Enter into integrated agreements for socio-economic development
- Establish education and training partnerships
- Establish employment development partnerships
- Establish business development partnerships
- Establish joint environmental initiatives
- Enter into collaborative resource management initiatives

Useful First Nations & Mining Contact Information

Aggregate Producers Association of British Columbia
http://www.gravelbc.ca

Assembly of First Nations
http://www.afn.ca/

BC Government First Nations Consultation and Other Information
http://www.mcaws.gov.bc.ca/aboriginal_dir/guide.htm

BC & Yukon Chamber of Mines
http://www.chamberofmines.bc.ca

Canadian Aboriginal Minerals Association
http://www.aboriginalminerals.com

First Nations Summit
http://www.fns.bc.ca

Indian and Northern Affairs Canada
http://pse-esd.ainc-inac.gc.ca/esd-pse/index_all_e.asp

Indian Resource Council of Canada
http://www.indianresourcecouncil.ca

Mining Association of BC
http://www.mining.bc.ca/

Mining Association of Canada
http://www.mining.ca/

National Aboriginal Business Association
Tel: 604-275-6670; 1-800-337-7743

Native Investment and Trade Association
http://www.native-invest-trade.com

Northeast Aboriginal Business Centre
http://www.aboriginalbusinesscentre.com/

Prospectors and Developers Association of Canada
http://www.pdac.ca/

Union of BC Indian Chiefs
http://www.ubcic.bc.ca/

NOTES

NOTES